SIGNAL FROM SPACE

by Will Eisner

Princeton KITCHEN SINK PRESS Wisconsin

I would like to give special thanks to my old colleague of Spirit days, Andre LeBlanc, who assisted on the coloring of this book. —Will Eisner

ISBN No. 0-87816-014-0

Printed in Italy

Cover paintings by Will Eisner. Logo design by Peter Poplaski. Project editor: Jens Peder Agger

 Published by Kitchen Sink Press, Inc., No.2 Swamp Road, Princeton, Wisconsin 54968 USA. The material in this collection originally appeared in *Will Eisner's Spirit Magazine*, serialized under the title "Life On Another Planet," in black and white.

CHAPTER 1
THE SIGNAL
LIFE ON ANOTHER PLANET
BY Will Eisner

A message etched on metal has been placed aboard Pioneer Ten & Eleven by NASA and sent out into space where perhaps some distant intelligence, with the means of reaching us will respond . . . or at least will know that they are not alone in the universe.

INTRODUCTION

In 1950, the National Radio Astronomy Observatory at Greenbank, West Virginia set up an attempt at listening to radio emissions from space. Their hope was to intercept signals from distant civilizations in our galaxy which would indicate **intelligent life on another planet.**

In the twenty years that followed, Canada and the Soviet Union made similar attempts – no one achieved tangible results. By 1978 almost 1000 stars had been examined by U.S. scientists alone. Nearby stars like, epsilon fridan and tau ceti got special attention.

THE RESULTS HAVE BEEN NEGATIVE!

Nevertheless, the National Aeronautics and Space Administration continues the effort in various other installations around the United States. In the Soviet Union (the Caucasus) a large 600-ratan radio telescope devotes some of it's activity to a similar search for signs of life on another planet.

Stimulated by the mathematical probabilities some modern scientists have this hypothesis: making "guestimates" based on a system that calculates the age of stars and their planets, and assuming only one star in any group of 250,000 (out of the billions that exist) has a planet that can (and does) support intelligent life – there are enough planets within our reach to justify a very reasonable possibility of this. Also, we have the technology to make contact.

As early as May 30, 1916 Bernard's star was noticed by astronomers to have unusually rapid motion. In recent years signals from it have caught the attention of modern space probers because of their unusual nature.

SOMETIME BETWEEN MIDNIGHT AND DAWN - ON THE MESA RADIO ASTRONOMY OBSERVATORY, NEW MEXICO...
?!
CHAK CHAK CHAK
CHAKCHAKCHAK
CRYPTO DEPT.
WAKE UP MALLEY... I HAVE SOMETHING THAT WILL BLOW YOUR MIND!
HUH?
CLIK

I WANT YOU TO GET INTO YOUR COMPUTER AND DECODE THESE BLIPS, NOW!!
YAWN ...THIS IS SOME KINDA CREEPY CODE!
MAN... THIS IS LIKE GIBBERISH... IT'S SIMPLE...YET, WEIRD!
AHA ...SEE? THEY'RE IN A KIND OF MODULATED SEQUENCE-OBVIOUSLY COMPRISING THE NUMBERS 1, 3, 5, 7, 11, 13, 19, 33, 29, 31... THEY CAN BE DIVIDED BY ONE.. OR BY THEMSELVES!
SO?!
SO, THIS IS A SIMPLE MATHEMATICAL TYPE CONCEPT... IT DOESN'T SAY ANYTHING! IT JUST DOES THE ARITHMETIC... OVER-AND-OVER!
WHAT'S THE MATTER ARGANO- YOU LOOK SICK!?
OH MY GOD! THAT MEANS IT'S... A SIGNAL! IT COULD ONLY HAVE A BIOLOGICAL ORIGIN... A...A FORM OF INTELLIGENCE... LIKE OUR OWN... FROM OUTER SPACE!

THIS IS A GAG! YOU GOTTA BE KIDDING!
I'LL SHOW YOU...
SEE, IT'S STILL COMING IN...
WOW!
NOW, IT'S STOPPED!
THEY'VE STOPPED SENDING! OBVIOUSLY WAITING FOR A RESPONSE!
I'LL REPORT THIS...
NO!
THINK, MALLEY... THINK!! THINK OF THE POSSIBILITIES...
LET GO!
LOOK, ARGANO- WE JUST WORK HERE... WE CAN'T CRAP AROUND!
PLEASE, MALLEY... KEEP YOUR MOUTH SHUT A FEW HOURS! I NEED A BIT OF TIME TO THINK!

5/1

WHY ARE WE GOING TO TRACKING?
WE'VE GOT TO GET A FIX ON THE SOURCE!
TRACKING
EVENIN', I'M DR. MARK ARGANO OF MONITORING - THIS IS PROF. MALLEY OF CRYPTOLOGY... GOT A MINUTE MR. COBBS?
SURE, WHAT'S COOKING?
COULD YOU TRACK... AND GIVE US THE SOURCE OF THIS SIGNAL?
MM... LOOKS EASY.
BASED ON STRENGTH AND FREQUENCY I FIX IT SOMEWHERE IN THIS QUADRANT! WHAT'S THE DEAL?
OH, ...ER JUST CURIOUS.
YEH, YEH!
BULL!! LET ME IN ON IT... OR NO MORE HELP FROM ME!
OKAY... OKAY... WE GOT A SIGNAL... DECODE SAYS ITS A KIND OF MESSAGE.
THERE'S AN INTELLIGENT LIFE ON ANOTHER PLANET TRYING TO REACH US.. WE NEED TO FIND THE PLANET

OKAY... HERE...I MAKE IT A PLANET OF THE STAR BERNARD, ABOUT TEN YEARS AWAY-AT CURRENT SPACE TRAVEL CAPACITY...IT'S A TINY PLANET-NOT BIGGER THAN OUR OWN MOON! ...THAT'S THE BEST I CAN DO FOR NOW.
GOOD WORK COBBS
OKAY, NOW-WE ARE THE ONLY PEOPLE ON EARTH WHO KNOW ABOUT THIS! I WANT YOU BOTH TO SWEAR YOU'LL KEEP IT A SECRET UNTIL I CAN FIGURE OUT HOW TO CAPITALIZE ON IT!
OKAY, I SWEAR!
YEAH, SURE, WHY NOT!
WASHINGTON, D.C. 9:30 P.M. E.S.T.
COMRADE ZORKY...IT'S OUR 'MOLE' IN NEW MEXICO WITH A CODE 6... HIS NAME IS COBBS
CONSULATE USSR
I CAN'T TALK NOW COMRADE... I ASSURE YOU I WOULD NOT RISK THIS CALL IF IT WERE NOT VERY IMPORTANT!

...YES,SIR, I'M AT HOME ALONE...I'LL WAIT HERE UNTIL YOUR LOCAL CONTACTS ME!
GAWD HE'S A SPY! 'GOOD THING WE CAME OUT HERE ARGANO!
SHH
CAREFUL, MALLEY... HE'S GOT A GUN!
HOLD IT!
YOU'RE A LOUSY COMMIE AGENT, COBBS
WE'RE GOING TO TURN YOU IN!
NOW, COOL IT... YOU GUYS HAVE BEEN SITTING ON THIS STUFF TOO LONG- THEY'LL SUSPECT YOU'RE IN IT WITH ME... SEE!
IT'S A LIE WE'RE NOT SPIES

LOOK AT IT THIS WAY... THIS IS INFORMATION THAT BELONGS TO THE WORLD -NOT FOR SOME HUSTLER ...OR ONE LOUSY CAPITALIST NATION... JUST THINK OF THE HISTORY OF MAN'S DEALINGS WITH NEW TERRITORIES... WE'RE BARBARIANS! NO, AT LEAST IF TWO MAJOR NATIONS SHARE IN IT THEY WILL HAVE SOME CHECK ON EACH OTHER!
...THIS IS A MORAL ISSUE! BESIDES - THAT SIGNAL WILL BE BEAMED AT EARTH AGAIN... THEN THE U.S. WILL HAVE IT TOO...SO, WHAT'S THE BIG DEAL?
LISTEN ARGANO...THE KGB DOESN'T YET KNOW WHAT I HAVE ...SO, WHEN THEY GET HERE I'LL SHARE CREDIT WITH YOU. THEY'LL ARRANGE FOR YOU TO DEFECT TO THE U.S.S.R...YOU'LL BE A BIG SHOT THERE!
HOW ABOUT THAT?
IT STINKS COBBS ...NO!
PLEASE ARGANO, LET'S THINK IT OVER ...MAYBE WE...
NO!

THE WHOLE DEAL IS MINE! I DON'T GIVE A DAMN ABOUT ANY MORALITY CRAP!
KEEP BACK ARGANO!
THIS IS MY CHANCE TO MAKE IT BIG...
NEVER MIND LOOKIN' FOR HIS GUN-GIMME THAT ROCK, MALLEY!
HELP ME GET RID OF THIS ...UGHHH
ULKKK I'M GOING TO VOMIT
NOT HERE MALLEY-LISTEN, HE'S A LOUSY SOVIET SPY...NO BIG LOSS! THEY'LL KEEP IT HUSH-HUSH! GO HOME. TOMORROW WE'LL REPORT TO WORK LIKE NOTHING HAPPENED!

WASHINGTON D.C.
10:40 P.M
EST
NU GROSKY ?
...COBBS WAS KILLED LAST NIGHT BEFORE OUR LOCAL COULD GET TO HIM!
CONSULATE USSR
...HMM... LOOKS LIKE HE MAY HAVE HAD SOME-THING! BIG!
CONSULATE USSR
GET OUT THERE... FIND OUT WHAT HE WAS GOING TO TELL US.
CONSULATE USSR
WASHINGTON D.C.
11:40 P.M
EST
RING RING
TELEPHON
RING RING
...HELLO?! BLUDD HERE...YES SIR, I'LL BE RIGHT THERE ...I UNDERSTAND!
TELEPHON
11.1

...WHY ALL THE CLOAK-AND-DAGGER... I'M AN ASTROPHYSICIST NOT A SPY! WHAT'S THE CIA WANT WITH ME, SAM?
SHUT UP BLUDD, AND LISTEN!!!
IT LAYS OUT LIKE THIS..... EARLY LAST NIGHT A TRACKER FOR THE MESA RADIO ASTRONOMY OBSERVATORY WAS KILLED. HE WAS A SOVIET PLANT! OUR MOLE IN THE RUSSIAN EMBASSY TELLS US THEIR PLANT WAS ABOUT TO DELIVER SOMETHING HOT!
YOUR JOB IS TO FIND OUT WHAT IT WAS!
BUT...
I'M JUST A MESSENGER BOY... HERE, TAKE THESE PLANE TICKETS AND GET OUT OF HERE - YOU WILL BLOW MY COVER IF YOU HANG AROUND!
NEW MEXICO
ONE WEEK LATER
IS THERE ANY THING MORE I CAN DO FOR YOU DR. BLUDD.. DO YOU WANT ME TO CLEAN OUT MR. COBBS' DESK, SIR?
AHH... NO, JUST LEAVE EVERYTHING AS IT IS! I'LL CLEAN UP MYSELF!
124

I FEEL A BIT UNEASY AS THE REPLACEMENT OF A DEAD MAN! WHAT HAPPENED TO COBBS, MISS BOWEN?
OH, THEY SAY HE WAS MUGGED- JUST OUTSIDE HIS HOUSE! THE POLICE SAY ROBBERY WAS THE MOTIVE!
AT ANY RATE, I'M YOUR NEW SECRETARY - CALL ME IF YOU NEED ANY HELP!
THANK YOU!
TWO WEEKS LATER
YOU WANTED TO MEET WITH US HERE IN MY HOUSE, DR. BLUDD? ...WHAT'S SO URGENT?
WELL, DR. ARGANO, AS YOU KNOW-I'M COBBS' REPLACEMENT. TWO DAYS AGO I FOUND SOME DISCARDED NOTES IN HIS DESK...
UH, OH!
SHUT UP MALLEY!
...THESE NOTES PLUS MY OWN CALCULATIONS-AND THESE TAPES OF RESIDUAL IMPULSES ON YOUR MONITORS LEAD ME TO THE CONCLUSION THAT YOU **RECEIVED A SIGNAL FROM SPACE!**

WHO THE HELL ARE YOU?
I'M DR. JIM BLUDD. ...ACTUALLY I'M AN INVESTIGATOR... ...YOU GUYS ARE IN DEEP TROUBLE
LOOK... WE'LL LEVEL WITH YOU... ER, AH.. Y'SEE COBBS WAS A SPY. HE WANTED TO PASS THAT INFO TO THE SOVIETS. WE HADDA STOP HIM!
C'OME ON BLUDD GIVE US A BREAK!
WE CAN BE OF HELP. ...THESE TAPES NEED INTERPRETATION BLUDD!
HMM.. WE'LL SEE...
NOT NECESSARY BLUDD... WE'LL TAKE THOSE TAPES NOW!
WHY... MISS BOWEN, YOU'RE A SPY... TSK TSK...
OH, STOW IT... THE TAPES PLEASE-WE ARE IN A HURRY!
TAKE **THOSE TWO** ... WE DON'T NEED BLUDD!
14/1

MISS BOWEN
I
KEEP IT CLEAN! LEAVE BLUDD IN THE HOUSE
IN THE CAR YOU TWO!
SETTLE DOWN DR. ARGANO, PROFESSOR MALLEY... WE ARE GOING ON A LONG TRIP!
15/1

WASHINGTON D.C
BOY, BLUDD HOW YOU GOT OUT OF THAT BURNING HOUSE... I'LL NEVER KNOW!! YOU WERE A MESS WHEN WE FOUND YOU!
WHERE'S ARGANO AND MALLEY ?
THEY WERE TAKEN TO MOSCOW-OUR MAN THERE SAYS THEY ARE TALKING. SO, THE OTHER SIDE KNOWS MORE THAN WE DO!
MY GOD !
TWO MAJOR NATIONS VYING TO MAKE CONTACT WITH AN UNKNOWN INTELLIGENCE ON A PLANET ONLY TEN YEARS AWAY FROM OURS... THINK OF IT...

CHAPTER 2
THE 1ST EMIGRE
GOOD EVENING, LADIES AND GENTLEMEN NASA, THE NATIONAL AERONAUTIC SPACE ADMINISTRATION TODAY ANNOUNCED THAT IT RECEIVED A SIGNAL FROM WHAT APPEARS TO BE INTELLIGENT LIFE ON THE PLANET BERNARD - A SMALL MOON ESTIMATED AT ABOUT 10 YEARS DISTANT FROM THE EARTH

THE REPORT, WHICH THE ADMINISTRATION HAS TRIED TO KEEP SECRET ON GROUNDS OF NATIONAL SECURITY HAS BEEN LEAKED TO THE PRESS CAUSING THE SPACE AGENCY TO FINALLY CONFIRM WHAT COULD BE THE MOST SIGNIFIGANT DEVELOPMENT IN THE HISTORY OF MANKIND. HERE IN THE GBC STUDIOS WE HAVE DR AMOS GAWLT, EMINENT ASTRONOMER, TO GIVE US SOME INSIGHT.
. . . . . .
DR. GAWLT, WHAT KIND OF SIGNAL DID WE RECIEVE?
. . . .
ACTUALLY, WHAT MESA, OUR TRACKING STATION PICKED UP WAS A SERIES OF MATHEMATICAL EQUATIONS...
©!#@!
ROTTEN LOUSY NO GOOD LIFE
2.2

WE ONLY ASSUME WE'RE GETTING A SIGNAL BECAUSE THEY WERE REPEATED OVER AND OVER ... THREE TIMES, I BELIEVE ...
GIVEN THE FACT THAT THIS IS A SIGNAL, THEN WE CAN CONCLUDE THERE IS AN INTELLIGENCE - INDEED, SOME INTELLIGENT LIFE TRYING TO COMMUNICATE WITH US.
....
HMM.... THEN, DR. GAWLT, WHAT IS THE NEXT STEP?
......
WELL, ALL WE CAN DO AT THIS TIME IS SEND THAT SIGNAL BACK... THIS WILL TELL WHOEVER IS UP THERE THAT WE RECEIVED IT AND HOPEFULLY ESTABLISH A PRIMARY CONTACT.
......
BUT, DR. GAWLT, CAN'T WE SEND A SPACE PROBE UP THERE?
GIMME A DRINK, JOE!
BEAT IT, MARCO... I'M LISTENIN' TO THE T.V... NO MORE DRINKS... Y'OWE ME 50 BUCKS ALREADY! BEAT IT!!
!
OH, YOU POOR MAN JOE SHOULDN'T'VE HIT YOU!
HERE... LET ME HELP YOU... C'MON MARCO I'LL TAKE YOU HOME.

WE HAVE PROBED MARS AND VENUS - WHY NOT THIS ?
NOT QUITE! YOU SEE, THE STAR BERNARD IS ABOUT 10 YEARS DISTANT BY OUR PRESENT TECHNOLOGY. ASSUMING WE COULD SEND A LAND AND RETURN VEHICLE OF, SAY, THE TYPE USED ON THE MOON, IT WOULD BE ABOUT 20 YEARS BEFORE IT RETURNED TO EARTH. EVEN SO, THE IDEA OF RISKING HUMANS IN SUCH A PROBE IS QUITE UNTHINKABLE.
THANK YOU, DR. GAWLT. WE HAVE ALSO, IN OUR STUDIO, AN INTERNATIONALLY FAMOUS THEOLOGIAN - DR. MARTIN BLUTH...
THERE, NOW, MARCO... LIE DOWN... I'LL GET YOU SOME COFFEE.
OH, DEAR WHAT A MESS... AND ALL THESE PAPERS.. YOU'RE A WRITER??
OOOHH MY HEAD... WHO IS THAT TALKIN'?? WHAT'S HE SAYIN?
?
THIS IS IT!! THIS IS THE WAY OUT... I WILL LEAVE THIS PUTRID PLANET
?
I, MARCO, WILL LEAD YOU TO THE PROMISED LAND! YES... THIS IS THE CALL... WE MUST GET READY!

DR. BLUTH... AS A CLERGYMAN WHAT DO YOU THINK THE RELIGIOUS SIGNIFICANCE OF THIS DEVELOPMENT WILL BE?
WELL, IT IS RATHER HARD TO GIVE A SIMPLE ANSWER TO THAT QUESTION. I DON'T SEE HOW A DISCOVERY OF AN INTELLIGENT LIFE FORM ON ANOTHER PLANET CAN DO ANY MORE THAN REINFORCE OUR BELIEF ... OUR FAITH IN GOD. AFTER ALL, THESE CREATURES -WHATEVER THEIR SHAPE - ARE STILL CHILDREN OF THE ALMIGHTY. THEIR AWARENESS OF GOD, AND HOW OR IF THEY WORSHIP IS ANOTHER MATTER.
I MUST TELL THEM I AM COMING!
BUT, HOW MARCO?
I HAVE A SHORT WAVE RADIO.. YES, YES... I WILL SEND A MESSAGE!
?
HELLO.. HELLO EARTH CALLING HELLO
IMAGINE... GOING TO ANOTHER PLANET... A NEW LIFE... A WHOLE NEW LIFE.. YES, YES
I'LL HELP YOU, MARCO! ...I'LL BE RIGHT BACK!
HELLO... EARTH CALLING

5/2

WELL, WE ARE NOW AT THE END OF A LONG DAY OF BROADCASTING AND COVERAGE OF THE SPACE AGE'S MOST STARTLING NEWS... WE NOW TAKE YOU TO WASHINGTON, D.C. FOR A STATEMENT BY THE WHITE HOUSE, LADIES AND GENTLEMEN, THE PRESIDENT OF THE UNITED STATES...

. . . .

GOOD EVENING, MY FELLOW AMERICANS. AT A MEETING OF OUR NATIONAL SECURITY ADVISORS IN MY OFFICE TODAY IT WAS DECIDED TO IMPLEMENT A PROGRAM WHICH WOULD RESPOND TO THE RECENT SIGNAL FROM SPACE.

I MUST POINT OUT, HOWEVER, THAT WE CANNOT HOPE FOR ANY SHORT TERM ANSWERS TO THIS VERY CHALLENGING DEVELOPMENT. INDEED, ANY RESPONSE TO PROBING BY OUR SCIENTISTS MIGHT WELL BE MANY YEARS AWAY. THERE IS NO NEED FOR SPECULATION OR OTHER ACTION THAN THAT WHICH NASA HAS UNDERTAKEN. I HAVE ASKED THE OTHER GREAT NATIONS, PARTICULARLY THE SOVIET UNION TO JOIN US IN A UNITED PROGRAM TO EXTEND OVER THE NEXT TEN YEARS. MEANWHILE, I URGE YOU, MY FELLOW CITIZENS, TO PROCEED WITH OUR COMMON EFFORT TO SOLVE INTERNATIONAL AND DOMESTIC PROBLEMS. GOOD EVENING TO YOU ALL.
AHA... CORA LISTEN...I'M GETTING A REPLY...
GOT A DRINK MARCO?
QUIET GIBNEY... HE SENT A MESSAGE WHILE I WAS GONE !!
LISTEN!
MY GOD, IT'S A VOICE... A VOICE FROM SPACE!
COME UNTO ME MY CHILDREN
NAHH IT'S JUST A SHORT WAVE SIGNAL FROM A CHURCH SERVICE IN AUSTRALIA, HAHA
GIBNEY.. YOU MUST BELIEVE BELIEVE !!!
OKAY, OKAY MARCO... LEGGO
GO FORTH SPREAD THE WORD.
MAN, HE'S FLAKY.
COME ON... ALL OF YOU ...WE MUST TELL THE WORLD.

MULTINATIONAL CORPORATION
STAR PEOPLE
JOIN
♫ JOIN THE STAR PEOPLE ♫ WE ARE THE FIRST TO EMIGRATE TO THE NEW STAR... THEY HAVE AREADY SENT US THEIR ♫ SIGNAL .. JOIN US
KOOKS
CRAZIES
GENTLEMEN, HAVE YOU READ THE CONFIDENTIAL MEMO ON THE RECENT SPACE SIGNAL?
IT SHOULD HEAT UP THE U.S. SPACE PROGRAM MULTINATIONAL SHOULD GET A GOOD SHARE OF IT!
AAH WASHINGTON'S PROGRAM OF A JOINT STUDY IS JUST TOO DAMN SLOW... IT'LL GET NOWHERE!
EXACTLY!.. IF WE CAN GET THE RUSSIANS MOVING UNILATERALLY ON THIS... THE U.S. WILL HAVE TO MOVE!
WE MUST ACCELERATE THIS THING!
STIMULATE COMPETITION ... SOVIET COMPETITION ??
BUT, MACREDY, WHAT OF AMERICA'S INTEREST.. THERE IS OUR RESPONSIBILITY YOU KNOW!

NEED I REMIND YOU, HARLEY, MULTINATIONAL IS A GOVERNMENT IN ITSELF... THROUGH OUR SUBSIDIARIES WE INFLUENCE THE ECONOMIC LIFE OF 15 NATIONS ON THIS EARTH... WE OWE THEM A HIGHER RESPONSIBILITY. WE ARE ABOVE THE POLITICS OF ANY ONE COUNTRY!
THEN IT IS THE SENSE OF THIS BOARD THAT WE PUSH THE USSR AND THE U.S. INTO A SPACE RACE!
BY WHATEVER MEANS, HARLEY!
I SEE!
OH, MR MACREDY, WE HAVE A RELIABLE REPORT THAT THE TWO AMERICANS, ARGANO AND MALLEY WHO FIRST HEARD THE SPACE SIGNALS, ARE IN SOVIET HANDS
GET THEM OUT! WE'LL NEED THEM... USE SENATOR BLITZ!
CLIK
MOVE OUT HUBBARD! I'VE GOT A DINNER MEETING...!! WHATS GOING ON?
SORRY, MR MACREDY THOSE KOOKY STAR PEOPLE ARE BLOCKING THE STREET! GOTTA WAIT 'TILL THE COPS CLEAR IT!

OH PLEASE... MISTER, LEMME IN... THEY'LL BUST ME...
SURE... JUMP IN.. HERE MISS! PULL OUT QUICKLY, HUBBARD!
THE FOOLS! WE ARE ONLY TRYING TO RECRUIT FOLLOWERS!
WHAT IS THIS STAR PEOPLE THING MISS?
WE ARE GOING TO THE PLANET BERNARD - OUR LEADER, MARCO WILL TAKE US!
A SIGNAL HAS ALREADY COME... WE'LL POPULATE THE NEW PLANET AN CREATE A NEW AND GOOD WORLD. !!
OUR NUMBERS GROW... LAST WEEK WE WERE FIVE... NOW WE ARE FIVE HUNDRED!
HUBBARD, TAKE THIS GIRL TO HER HOUSE... I'LL GET OUT HERE! GOOD BY, MISS, AND GOOD LUCK!
10.2

JONES?? MACREDY HERE!... I WANT YOU TO START A FLOW OF FUNDS TO THE STAR PEOPLE GROUP! ... DO AS I TELL YOU ... USE THE GORT FOUNDATION FUNDS TO FRONT IT!
I WANT THAT GROUP TO FLOURISH ... HAVE GREBE INFILTRATE THEIR MEMBERSHIP! ... YOU KNOW HOW TO HANDLE THAT! RIGHT!
... BEAUTIFUL, BEAUTIFUL!
WHERE TO MISS?
14th STREET
HARLEY,... ABOUT THAT MANNED ION-PROPELLED SPACE PROBE WE DEVELOPED IN OUR ENGLISH SUBSIDIARY... GET IT OFF THE SHELF AND INTO PRODUCTION!
BUT MR. MACREDY ...?
YOU REJECTED IT ONCE ... YOU SAID ONLY IDIOTS WOULD SIGN ON A MANNED DEEP SPACE PROBE SHIP ... IT WOULD BE A SUICIDAL JOURNEY!
WELL, NOW I'VE GOT THE IDIOTS ... BY THE TIME THAT SHIP IS BUILT, I'LL HAVE HUNDREDS OF EAGER ... ER VOYAGERS.

CENTRAL INTELLIGENCE
WASHINGTON, D.C.
WHAT? ME? I'M NOT GOING INTO RUSSIA TO GET ARGANO AND MALLEY... THAT'S INSANE! I'M AN ASTROPHYSICIST NOT A SPOOK!
BLUDD, YOU WERE THE LAST TO SEE THEM IN NEW MEXICO... YOU'RE THE ONLY ONE WHO COULD READ THOSE TAPES ...BESIDES, THEY MIGHT TRUST YOU!
...AND IT'S YOUR PATRIOTIC DUTY, SON!
WHO IS HE?
THIS IS SENATOR BLITZ OF THE ARMED SERVICES COMMITTE.
MY BOY... IT IS CRITICAL TO OUR NATIONAL SECURITY THAT WE HAVE THOSE MEN BACK HERE... AND TO KNOW WHAT THE SOVIETS HAVE LEARNED!
I'LL ARRANGE FOR YOU TO BECOME AN EMPLOYEE OF THE MULTINATIONAL CORPORATION. THEY HAVE A WEST GERMAN SUBSIDIARY. YOU CAN POSE AS AN ORDINARY RESEARCHER.
ACTUALLY, BLUDD...YOU DON'T HAVE TO GO INTO RUSSIA. OUR PROFESSIONALS WILL BRING THEM TO BERLIN. YOU, THEN, PERSUADE THEM TO COME BACK WITH YOU... SIMPLE?
12.2

DON'T WORRY BLUDD... WE'LL GIVE YOU COVER... BESIDES THOSE SOVIET AGENTS LEFT YOU FOR DEAD IN THAT FIRE BACK IN NEW MEXICO... IT'S A PIECE OF CAKE!
TWO DAYS LATER... NEW YORK
THIS IS MR. BLUDD, A NEW MEMBER OF MULTINATIONAL... HE'LL BE PART OF YOUR TEAM IN GERMANY... YOU'LL BE HIS AIDE TOMKINS!
AHH WELCOME ABOARD! ...YOU'LL LIKE WORKING FOR OLD MULTI!
MULTI
WELL, BLUDD WE'RE OFF FOR BERLIN! YOU'LL LIKE IT...GOOD FOOD, NIGHT LIFE... I KNOW MY WAY AROUND TOWN.
SOUNDS GOOD ...SO FAR, TOMKINS.
JOIN US JOIN THE STAR PEOPLE
JOIN THE STAR PEOPLE
TAXI

YES MARCO ?
SISTER CORA WE HAVE MARVELOUS NEWS .... COME HERE!
MEET MR GREBE A NEW FOLLOWER..TRULY, A BEAUTIFUL PERSON LET US WELOME HIM! HE BELIEVES
?
REALLY ??! WHERE DO YOU COME FROM BROTHER?
OH, SISTER CORA...I'VE SPENT MY LIFE IN THE MEANINGLESS ACCUMULATION OF MONEY.. NOW, PERHAPS I CAN FIND A NEW LIFE WITH ALL OF YOU ON A NEW PLANET!
JOIN THE STAR PEOPLE
HERE, MARCO, HERE FROM MY STORE OF WEALTH..$250,000. AND I SHALL GIVE YOU MY HOUSE ON 93rd STREET. IT HAS 20 ROOMS...IT WILL HOUSE US ALL! I WANT ONLY TO BELONG...TO SERVE....

BLESS YOU BROTHER ...WE ACCEPT YOUR OFFERING. YOU'RE ONE OF US NOW!
GIBNEY, TAKE OUR NEW BROTHER AND TEACH HIM OUR WAYS.
C'MON
HOLD IT GREBE... I SEEN YOU UP IN THE MULTINATIONAL BUILDING LAST WEEK... WHAT ARE YOU, A SPY?
SHHH...I CAN EXPLAIN, COOL IT!
HEY LEGGOOOO...
OH MY HELP, BROTHER GIBNEY COMMITTED SUICIDE, COME QUICKLY MARCO
15.2

HMM.. NO GREAT LOSS.. GIBNEY WAS NEVER A TRUE BELIEVER!
THERE IS NO TIME FOR GRIEF! GREBE...YOU WILL BE OUR NEW TREASURER
I SHALL SERVE YOU WELL MARCO
... AND NOW WITH OUR NEW FRIEND AND HIS WEALTH WE SHALL MULTIPLY... YOU, DEAR SISTER CORA WILL BE AT MY SIDE ALWAYS... TOGETHER WE BEGIN OUR VOYAGE TO THE STAR...
16.2

LIFE ON ANOTHER PLANET 3
A NEW FORM OF LIFE
NEW FEDERAL AGENCY NAMED TO PROBE SPACE MESSAGE
The U.S. Government has officially acknowledged today that the brief message received early this year from an unknown planet in space is genuine.
Accordingly, it has set up an ancillary agency to NASA for the purpose of furthering our attempt to make contact with what appears to be intelligent life on a distant planet.
In a speech at the Smithsonian the President stated that the existence of extraterrestrial life is a political matter rather than a scientific one and must be very carefully pursued.
Ever since the New Mexico Mesa tracking station first received the series of simple mathematical pulses, there has been a rising groundswell of interest.

'EVENING DR. CROWBEN... WORKIN' LATE?

DAMN DAMN!!
DR. CROWBEN WHAT'S THE MATTER?...
WE'VE NEVER SEEN YOU SO UPSET!!
READ IT YOURSELF!!
OH MY GOD!!
IT'S THE LAB REPORT ON YOUR PHYSICAL ...... THEY'RE NUTS!
NO... IT'S QUITE ACCURATE... I'M DYING GENTLEMEN... I'LL NOT LAST A FORTNIGHT... I'M DYING, DYING!!
3.

LISTEN,... I HAVEN'T MUCH TIME!! OUR WORK ON THE DNA AND CELLULAR STRUCTURE WILL NOW TAKE A NEW TURN! READ THIS CLIPPING!
...IT'S ABOUT THAT SIGNAL FROM SPACE! SO, WHAT'S IT GOT TO DO WITH OUR WORK ??
DON'T YOU SEE?... SOONER OR LATER THEY WILL WANT TO POPULATE THAT PLANET... SEND **PEOPLE** THERE!
SO?..
THAT PLANET MAY NOT BE ABLE TO SUPPORT **OUR** KIND OF LIFE... THEY MIGHT HAVE INTELLIGENT LIFE, BUT **NOT** IN THE FORM **WE** CALL **HUMAN**!
AHH...I GET IT... LIKE THIS PLANT... IT'S LIFE, BUT QUITE DIFFERENT-WITH IT'S OWN CELL STRUCTURE !!
YEAH... SEE HOW IT BENDS TOWARD THE SUNLIGHT ... THERE IS INTELLIGENCE OF A SORT HERE!
EXACTLY!!... SO WE ARE GOING TO CREATE A **NEW CELL**...A MUTATION FROM MY BODY AND THAT PLANT....STOP GAWKING!!... MY **DEATH** WILL HAVE A **PURPOSE**!!... NOW LET'S GET TO WORK!....
4.

GERMANY
TURN OFF THE HEADLIGHTS ...I'LL USE MY FLASH !!
BLUDD!!?
TOMKINS?
I'VE GOT MALLEY AND ARGANO... I BROUGHT THEM OUT AS I PROMISED, BLUDD!!
OVER HERE, TOMKINS... WHERE WE CAN TALK!
5.

I THOUGHT YOU DIED IN THE FIRE BACK IN NEW MEXICO,,, HOW?
NEVERMIND... I'M HERE TO GET YOU AND WHAT YOU KNOW BACK TO THE STATES!
HOLD ON, BLUDD... WE GOT A GOOD DEAL GOING - THE SOVIETS GIVE US A GOOD LAB,,, STATUS...
WE'RE NOT ABOUT TO TRADE THAT FOR PRISON BACK IN THE STATES!
WE'LL MAKE A DEAL!
JUST TELL ME WHAT YOU'VE GIVEN THEM....
...EVERYTHING WE KNOW.... THE SIGNAL...THE FIX... THE FREQUENCY... THEY'RE SURE THERE'S INTELLIGENT LIFE OUT THERE AND THEY AIM TO BE THE FIRST TO MAKE CONTACT....
FOR OPENERS THEY'LL SEND BACK THE SAME SIGNAL!
BUT, IT'LL TAKE YEARS... HOW'RE THEY GOING TO CUT TRANSMISSION TIME?
EVER HEAR OF NUTRINOS ?
6.

IN ILLINOIS-EARLY '77 A TEAM OF PHYSICISTS MEASURED A STREAM OF CHARGED PROTONS COMING FROM THE FERMILAB'S ACCELLERATOR--- MILES AWAY!...
THEY DISCOVERED THAT THE PROTONS HAD PASSED RIGHT THROUGH EVERY HILL OR NATURAL OBSTACLE AT THE SPEED OF LIGHT...
AS THEY ENTER A WATER TOWER CONSTRUCTED TO TRAP AND MEASURE THEM THEY CHANGED INTO **NEUTRINOS**!!
CONCLUSION: NEUTRINOS CAN GO THROUGH **ANYTHING**... SPACE DEBRIS, ASTEROIDS, PLANETS ...AT THE SPEED OF LIGHT! THE RUSSIANS HAVE DECIDED TO USE THEM FOR COMMUNICATING WITH THE STAR...
DAMN!
BY GOD... THEY'RE WELL AHEAD OF US!
ARGANO! I'VE **GOT** TO GET A LOOK AT THOSE CALCULATIONS !!
THAT MEANS GOING INTO RUSSIA...
LISTEN, BLUDD... YOU DIDN'T BARGAIN FOR THIS ...IT'S DANGEROUS !!
I'M HOOKED NOW... THERE'S NO OTHER WAY!! **TOMKINS**!
7.

ALL MY NOTES ARE IN MY ROOM BACK IN MOSCOW!
OKAY... HERES HOW WE PLAY IT... ARGANO GOES WITH BLUDD... I KEEP MALLEY HERE -AS HOSTAGE !!
HOW DO I GET OUT AFTERWARD, TOMKINS?
I'LL COME GET YOU... I'LL WAIT FOR YOU IN THIS LITTLE TOWN NEAR THE BORDER, BLUDD!!
HERE'S A MAP ...SEE, THIS IS GORKY STREET ... I'LL GIVE YOU 72 HOURS TO GET IN, PHOTOGRAPH THE DATA AND GET BACK TO RENDESVOUS!
OKAY, ARGANO...IT'S YOUR BALL NOW.... ANYTHING HAPPENS TO BLUDD-I'LL WASTE YOUR PAL, MALLEY!
OKAY, TOMKINS... NO SWEAT,
LET'S GO, BLUDD MOVE SLOWLY... ONCE WE'RE PAST THE BORDER I'LL DO ALL THE TALKING-KEEP YOUR FACE DOWN...

8

INCREDIBLE!
CROWBEN WAS RIGHT! IT WORKS... WE'VE GOT A PLANT'S DNA ISOLATED NOW,... WE CAN PERFORM A SIMPLE MUTATION WITH HIS DNA AND IMPLANT IT IN A CELL!
MORRIS CROWBEN DEAD AT 71 PRIZEWINNER DISCOVERED CELL MUTATION
PERFECT PERFECT!!
NOW WE HAVE A PROBLEM!
RIGHT!- WE'LL NEED TO INCUBATE THE NEW CELL,.. WE'LL NEED A LAB AND A SECRET PLACE TO DO IT!
...AND THAT'LL TAKE BIG MONEY... A GRANT FROM SOMEONE WHO BELIEVES.... NOT EASY...
STAR PEOPLE DEMONSTRATE
CULT WEALTH TOO VAST?
I THINK I'VE GOT THE SOLUTION!
STAR PEOPLE WEALTHY- CULT GROUP GROWING

STAR PEOPLE INTENATIONAL
GREBE!! COME IN HERE!!
... THIS IS DR. HOAD AND DR. JONES... THEY'RE SCIENTISTS WHO WANT TO JOIN US!
REALLY ?
STAR
WHY ?
... WE ARE WORKING ON A CELLULAR MUTATION SO EXPERIMENTAL THAT NO GRANT AGENCY WILL FUND IT... - BUT WE THOUGHT YOU MIGHT ... BECAUSE IT COULD MAKE LIFE POSSIBLE ON ANOTHER PLANET. !!
HMM... YOU REALIZE YOU WILL HAVE TO SIGN-OVER ALL RESULTS TO US!
WELL... AH, WHAT DO YOU THINK, JONESY ??
- SURE ... WHY NOT?
I WILL NEED A SUMMARY OF YOUR PROJECT AND LAB NEEDS!
SURE... I ANTICIPATED THAT - HERE!
RIGHT... YOU WILL HAVE OUR ANSWER IN A FEW DAYS.
WHAT DO WE WANT TO GIVE THEM MONEY FOR WHEN I NEED A CADILLAC ?
NOW, MARCO - PLEASE TRUST GREBE ?
10.

A MR. GREBE TO SEE YOU, MR. MACREADY....
?
WHAT ARE YOU DOING HERE, GREBE ..I TOLD YOU NEVER TO...
THIS IS BIG, SIR!! BRIEFLY, I'VE GOT TWO SCIENTISTS WHO'VE DEVELOPED A CELLULAR MUTATION...A LIFE FORM THAT CAN BE TRANSPORTED TO ANOTHER PLANET...
HMMM...
YES...YES...
TAP TAP
HOPKINS...THIS IS MACREADY... SEE IF YOU CAN MAKE A LAB AVAILABLE TO GREBE ...AND FUND THE PROJECT...
YES, SIR! WE HAVE AN IDLE LAB IN UTICA
AH, I KNEW YOU WOULD SEE ITS POTENTIAL, SIR!!
I HAVE THE ANSWER I PROMISED YOU LAST WEEK ... YOU HAVE THE MONEY... NOW WHEN CAN YOU START HOAD?
RIGHT AWAY, MR. GREBE!
MAYBE I'LL SEE YOU LATER...
MAYBE...
JOIN NOW
TO A NEW WORLD

11.

KEEP AWAY FROM... THE WINDOW... BLUDD... THIS IS MOSCOW ANY ODD MOVEMENT WILL CAUSE SUSPICION...
OKAY, ARGANO... LET'S SEE THE PAPERS!
HERE WE'VE GOT ALL THE COMPUTATIONS DONE...
HURRY, FER'CHRIST SAKE... SOMEONE'S COMING!!
HERE'S YOUR CAMERA... I'LL KEEP THE FILM!
GET IN THE CLOSET!
DARLING!!
HOW'D THINGS GO AT THE BUREAU TODAY, NADIA?
NADIA ?
NADIA!

WAIT, NADIA....
BLUDD!!... SO YOU DIDN'T DIE IN THE FIRE!
OBVIOUSLY... SO, EX-MISS BOWEN IS ARGANO'S LOVER ... NEAT...
YOU,...YOU TRAITOR!! AFTER ALL I'VE DONE FOR YOU...I KEPT YOU OUT OF THE K.G.B.'S HANDS... I GOT YOU PRIVILEGES!...
PLEASE LISTEN!
DO YOU REALIZE WHAT THIS WILL DO TO ME... MY CAREER... MY LIFE!
QUIET, NADIA-QUIET!
I'M GOING TO CALL THE K.G.B.
NADIA QUIET! NADIA NADIA... NADIA..
GLAK
GET OUT OF HERE, BLUDD!
LOOKS LIKE BOTH OF US HAVE TO GET OUT-FAST!
13.

NO, I'M STAYING I'VE GOT TO STAY AND HELP NADIA -I LOVE HER!
HAVE IT YOUR WAY, ARGANO... SO LONG!
TOMKINS... THAT YOU?
WHEW... YOU SURE ARE DEPENDABLE!!
BOY, IT WAS HAIRY- BUT I HAVE THE FILM AND...
TOMKINS!!?
14.

OH MY GOD!
-GOT TO GET TO THE U.S. EMBASSY!
HOLD IT, BLUDD!
ARGANO... YOU FOLLOWED ME!!
I CAN'T LET YOU OUT, BLUDD,... YOU'RE THE ONLY CHANCE I HAVE TO SQUARE IT WITH NADIA!
OUT OF MY WAY!!...
I'M TAKING YOU BACK...
NO WAY, ARGANO... I'VE NEVER KILLED ANYONE IN MY LIFE... BUT IT'S YOU OR ME!!
15.

RING RING ... RING RING ...
SIR ,,, A MR. BLUDD JUST CAME INTO OUR EMBASSY IN MOSCOW ,,, BADLY BEAT UP, BUT ALIVE ,,, SAYS HE WORKS FOR US ... YOU KNOW HIM? ...
YES, GET HIM BACK – AND KEEP A LID ON HIM ! ,,, HE'S VALUABLE PROPERTY.
MR. MACREADY, OUR MR. BLUDD HAS JUST BEEN RETURNED BY THE CIA – BUT WE LOST TOMKINS !!
ULTINATIO
DID BLUDD BRING ANYTHING OUT !!
YES SIR, HE GOT WHAT WE SENT HIM FOR !!
BEAUTIFUL, BEAUTIFUL !
BEAUTIFUL.. LOOK AT THIS CULTURE, HOAD ...
JONESY ,,, WE'RE RIGHT ON ! IF THE INCUBATION WORKS WE'VE GOT A NEW FORM OF LIFE !!

16.

LIFE ON ANOTHER PLANET 4
PRE-LAUNCH
...WHAT A SIGHT... A SUPINE CITY CONCERNED WITH MERE **EXISTENCE**-WHILE A CIVILIZATION ON A **DISTANT PLANET** IS TRYING TO MAKE CONTACT-**AN IMPACT THAT COULD CHANGE THE DESTINY** OF HUMAN KIND!! YET, WE CRAWL ABOUT DOING LITTLE THINGS... ARE WE, AFTER ALL, ONLY **ANTS** ON A PIECE OF SPACE DEBRIS CALLED EARTH??
PLEASE WAIT HERE **MR. BLUDD.** A SECRETARY WILL ESCORT YOU.
MR. MACREADY WILL SEE YOU NOW, MR. **BLUDD.**
1.

THIS WAY MR. BLUDD
QUITE A SUITE OF OFFICES, MISS...
WELL, AFTER ALL... THIS IS THE TRIPARTATE COMMISSION!
CLICK CLACK CLICK CLACK CLACK CLICK CLICK CLACK CLACK CLACK CLICK CLICK
BANG BANG BANG BANG BANG BANG CLICK CLICK CLICK
JOIN THE STAR PEOPLE
EMIGRATE TO A NEW PLANET
Members Welcome
ONE FLIGHT UP
2.

MR. BLUDD FOR MR. MACREADY.
THIS WAY MR. BLUDD!
YOUR FINANCE MINISTER HAS A REPORT ETERNAL PRESIDENT AMI
SPEAK UP OTUBI ... DON'T BE AFRAID!!
OUR NATION IS BANKRUPT... THE TREASURY OF THE GREAT NATION OF SIDIAMI IS EMPTY!!
GET ME A NEW FINANCE MINISTER
WE NEED MONEY! I GOT AN ARMY TO PAY... A PALACE TO RUN... MY HORSES... MY CARS... TAX THE NORTHERN TRIBES MORE! YOUR JOB IS TO RAISE MONEY!
WE CAN'T!! WE DON'T PRODUCE ENOUGH FOR EXPORT... OUR PEOPLE DO NOT FARM... WE HAVE LITTLE TO SELL FOR FOREIGN EXCHANGE TO BUY MACHINES... WE OWE ONE HUNDRED BILLION TO OTHER NATIONS

THIS IS MR. MBOBE... YOUR NEW FINANCE MINISTER... ETERNAL PRESIDENT
INTELLIGENCE REPORT SIGNAL FROM PLANET BERNARD
CAN WE POSSIBLY PAY ON ANY DEBTS TO THE OTHER NATIONS MR. MBOBE ??
NO, WE OWE OVER 100 BILLION TO 5 NATIONS MOSTLY IN THE WEST...
WE SHALL SECEDE FROM THE PLANET EARTH ...WE WILL NO LONGER OWE MONEY TO ANYONE!
SO... THEN WE SHALL FREEZE ALL FOREIGN PROPERTY... CLOSE OUR BORDERS AND...
I DECLARE OUR MEMBERSHIP IN THE PLANET BERNARD! INVITE THE STAR PEOPLE CULT TO SIDIAMI!! WE ARE NOW A COLONY OF ANOTHER PLANET!!
JUST TAKE THE ELEVATOR TO THE FIRST STOP, MR. BLUDD
THANK YOU, MISS
CLICK CLACK CLICK CLACK CLACK CLICK CLACK
CLACK CLICK CLACK CLICK
4.

MR. MACREADY ?
AHH WELCOME MR. BLUDD... COME ON IN!
NEXT TO ME... YOU'LL BE SITTING IN ON A SPECIAL MEETING OF THE TRIPARTATE COMMISSION
WHAT'S THE TRIPARTATE COMMISSION?
AN INTERNATIONAL COMMITTEE OF LEADERS IN ALL FIELDS OF HUMAN AFFAIRS... QUITE INFLUENTIAL!! SHHH, WE'RE CONDUCTING A REVIEW...
... AND THESE ARE THE MAJOR WORLD-REPORTING POINTS RESPONDING TO OUR SURVEY...
WHAT IS THE GENERAL RESPONSE TO THE SPACE SIGNAL REPORT?
SURPRISINGLY INDIFFERENT!... OF THE 3000 CULTS IN NORTH AMERICA, FOR INSTANCE, ONLY A FEW HAVE DEALT WITH THE SUBJECT. NOTABLY, THE STAR PEOPLE, WHICH NOW NUMBERS 10,000 FOLLOWERS IS THE MAJOR GROUP! THEY ARE REPORTED TO COLLECT ABOUT $5 MILLION ANNUALLY!

THANK YOU, DR. COLE... AND WHAT ABOUT THE ESTABLISHED RELIGIOUS GROUPS?
OFFICIALLY... A WAIT-N'-SEE ATTITUDE PREVAILS. THERE ARE ABOUT 350 ARTICLES IN PUBLICATION ABOUT THE RELIGIOUS SIGNIFICANCE OF INTELLIGENT LIFE ON ANOTHER PLANET.
FINANCE ??
ECONOMIC PATTERNS ARE UNAFFECTED, MOSTLY -BUT SOME SMALL COMPANIES ARE PROMOTING SPACE RADIOS...TOYS!!
INDUSTRY?
...LITTLE IMPACT! OH, A MARKED RISE IN ABSENTEEISM - BUT HOW MUCH WE CAN ATTRIBUTE TO THIS IS QUESTIONABLE !!
...AND SCIENCE ??
A RASH OF ABSTRACT RESEARCH! ...UNIVERSITIES ARE ACTIVE. THE NATIONAL SCIENCE RESEARCH PEOPLE ARE MOST CONCERNED! THEY SEE OUR PLANET, EARTH, IN DIRE NEED OF ENERGY SCOURCES AND AGRICULTURE SPACE! IN THE NEXT CENTURY, THE CONSENSUS IS THAT WE MUST MAKE CONTACT WITH WHO-OR-WHATEVER IS ON THAT PLANET!
WHAT DO WE REALLY KNOW ABOUT THAT PLANET OR ITS SUN ?
WELL, FIRST OF ALL THE NATURE OF THE SIGNAL IMPLIES AN ADVANCED TECHNOLOGY... AT LEAST EQUAL TO THE TASK OF SENDING A SIGNAL!
OUR ASTRONOMIC DATA, ADMITTEDLY SLIM, PERMITS ONLY A GUESS! ...WE THINK IT MAY HAVE TROPICAL ATMOSPHERE, MOISTURE, HYDROGEN OR OTHER GASSES... WE KNOW LITTLE ELSE!

NOW, SIR... SUPPOSE WE COULD TRANSPORT AND IMPLANT LIFE ON THAT PLANET TO INSURE OUR PRIMACY??
—NONSENSE!! WE CAN'T RISK A HUMAN LIFE ON SO FRIVILOUS AN IDEA!
NOT HUMAN LIFE AS WE KNOW IT—BUT A LIFE-FORM FABRICATED FOR THE PURPOSE!
DR. JONES... YOUR REPORT ... PLEASE!
WE HAVE SUCCEEDED IN CREATING A CELLULAR MUTATION... A HUMAN CELL-WITH-A-PLANT CELL TO FORM AN ORGANISM THAT CAN SURVIVE ON A PLANET LIKE BERNARD, AND...
...WE BELIEVE THAT WITH THE INSEMNATION OF PEPTIDES, SUCH AS THE ENDORPHIN (Substance P) ONE OF THE CHEMICALS FOUND IN HUMANS WHICH PROVIDES REACTION TO DISTRESS LIKE; PAIN, PLEASURE ETC... AND VASOPRESYN- A PEPTIDE WHICH FUNCTIONS AS A PROTECTION OF MEMORY—ALSO, BY ADDING BRADYKININ AND NEUROTENSIN (WHICH IN HUMANS EFFECT BODY SECRETION AND TEMPERATURE) WE CAN INDEED SIMULATE A LIFE-FORM SUITABLE TO YOUR PURPOSE.
GOOD LORD!
THIS IS SCIENCE FICTION!
GASP!
GASP
NONSENSE
GENTLEMEN... GENTLEMEN!! YOU ARE MISTAKEN...THIS IS JUST AN EXPEDIENT...WE ARE JUST ACTING PROMPTLY UNTIL HUMAN BEINGS CAN MAKE THE JOURNEY IT IS ECONOMIC FORESIGHT ... PURE AND SIMPLE!
WE ARE OUT OF OUR AREA OF COMPETENCE! ...IT IS A NATIONAL RESPONSIBILITY NOT OURS...
7.

On the morning of February 2, 1966, one of America's foremost lie-detector experts made a fascinating discovery. The man's name was CLEVE BACKSTER; the strange revelation—which was to initiate a continuing series of the most unusual experiments in the history of horticulture—began when he decided to find out how long it would take water to travel from the roots to the leaves of an office plant. To satisfy his curiosity, he attached a pair of polygraph electrodes to a leaf of his *dracena massangeana*. The result stunned him. So much so that it has altered his life from that day.

What Backster expected to find was a decrease in electrical resistance, induced by increased moisture. Instead, to his astonishment, he observed a psychogalvanic reflex (PGR) pattern closely resembling that of a human under emotional stimulation!

Mr. Backster was far from a novice in the use of the polygraph, an instrument employed to detect emotional changes in humans and, thus, betray their true reactions to questioning.

Mr. Backster had, in fact, served as a C.I.A. polygraph section head for a number of years, and regularly taught police, private investigators and government personnel how to use it. An acknowledged expert, he was not likely to be mistaken about what he saw: his plant was apparently displaying *emotion!* To verify the amazing P.G.R. reading, he tried a number of experiments. First, he dropped brine shrimp into boiling water, in the presence of plants. The P.G.R. reading showed violent reaction. It appeared that some sort of extrasensory broadcast of danger was being transmitted to the plants by the dying shrimp cells. He repeated the killing experiment with other forms of life—fresh fruit and vegetables, yeast, mold, amoebae, paramecia, sperm—and, each time, the plants recorded the same violent reaction on the polygraph! Backster concluded that killings in the immediate vicinity of the plants proved that they share a characteristic with animals: each stakes out a territory and reacts to phenomena occurring within that space.

Further experimentation revealed that plants seem to possess a sort of "adaptive logic" or "memory," since repeated killings in their territory produced a diminishing reaction. This demonstrated, he concluded, that the subject plants gradually perceived that the killing of the shrimp did not endanger *them*. Incredible as it may seem, this indicated that plants can *THINK!*

As the result of several thousand tests of polygraphed plants since 1966, Backster has made the following observations:

PLANTS REACT SELF-DEFENSIVELY! When they feel threatened, they respond like animals or humans. According to P.G.R. readings, they can actually faint!

PLANTS RECOGNIZE HUMAN ENEMIES! They appear capable of loving, hating or fearing people. In one experiment, a physiologist (who customarily destroyed her plants when her testing was completed) entered Backster's laboratory and his plants "froze up" with fear. When she departed, they relaxed.

PLANTS DEVELOP RELATIONSHIPS WITH THEIR MASTERS! Monitoring showed that Backster's plants continued to react to his thoughts even when he moved to another room, or a distant building. Using automated equipment and precisely synchronized stopwatches, he proved that his plants registered an emotional reaction at the exact moment he *decided* to return from New Jersey to his New York lab . . . a distance of about 15 miles. Further tests demonstrated that no shield, Faraday Cage, lead container or similar enclosure could block out or interfere with the extrasensory communication link between plant and master.

THAT BRINGS US AT LAST TO THE POLITICAL ASPECTS OF THIS REVIEW! ...IF WE ARE THE FIRST TO POPULATE THAT PLANET WITH OUR LIFE FORM WE INSURE WESTERN DOMINANCE IN ANY FUTURE COMMERCIAL EXPLOITATION!
MR. BLUDD, A CIA AGENT AND AN ASTROPHYSICIST, HAS JUST RETURNED FROM THE SOVIET UNION... WHAT IS THE SITUATION THERE JIM??
THE USSR HAS ALREADY SENT A RESPONSE SIGNAL TO THE PLANET VIA BEAMED NUTRINOS.
AHEM... EXCUSE ME..MR. MACREADY
GENTLEMEN... I HAVE A NEWS FLASH... THE COUNTRY OF SIDIAMI HAS SECEEDED FROM THE FAMILY OF NATIONS... IT PROCLAIMS ITSELF A COLONY OF THE PLANET BERNARD... A FORTUNATE DEVELOPMENT!
THEY'RE CRAZY!
HA HA HA HA HA HA HA HA HA HA HA HA HA
...NOT SO CRAZY! I PROPOSE WE ESTABLISH A LAUNCHING COMPLEX IN THAT COUNTRY AND SEND OUR 'IMMIGRANTS' OFF FROM THAT COUNTRY!
THE U.S. GOVERNMENT WILL NEVER SUPPORT US! ...NOR WOULD ANY OTHER NATION
THE POLITICIANS WILL CHEW A PROJECT LIKE THIS TO BITS
OBVIOUSLY! THAT'S WHY SIDIAMI IS THE PERFECT LAUNCHING PAD!
THIS IS BIGGER THAN PETTY POLITICS.... MULTINATIONAL CORPORATION WILL PROVIDE THE TECHNICAL KNOW-HOW!! IT IS AFTER ALL, A COMMERCIAL ENTERPRISE
9

... AND **MR. BLUDD** WILL DIRECT THE OPERATION
JUST A MOMENT! ...**HOLD ON!**
A PROBLEM MR. BLUDD ?!
THIS IS **INSANE**... IT'S **IMMORAL!** ...TO GAIN A COMMERCIAL ADVANTAGE YOU ARE PUSHING OUR SOCIETY INTO AN **ADVENTURE** THAT WILL INFLUENCE THE **FUTURE** OF **LIFE ON EARTH WITH NO CONCERN FOR THE CONSEQUENCES!**
... NOT A MAN IN THIS ROOM EVEN QUESTIONED THE MORALITY— OF CREATING A NEW LIFE FORM!
COUNT **ME OUT** GENTLEMEN !!
SLAM
TAP TAP TAP
IT APPEARS WE SHALL HAVE TO PROCEED **WITHOUT MR. BLUDD!**
10

NO! NO! NO! DON'T BE NAIVE, BLUDD... Y'DON'T WALK IN HERE AND GET AN APPOINTMENT WITH THE PRESIDENT!
CENTRAL INT
THIS IS NOT AN ORDINARY MATTER!
TICK TICK TICK
I'LL SEND A MEMO UP THRU CHANNELS... BEST I CAN DO!!
NOT GOOD ENOUGH ...MULTINATIONAL CAN ONLY BE STOPPED BY PRESIDENTIAL ACTION—FORCEFUL AND IMMEDIATE!
TICK TICK
NOT MUCH I CAN DO ABOUT THAT, BLUDD!
... WELL, I CAN... I CAN LEAK IT TO THE PRESS!!
TICK TICK TICK
TICK TIC
TICK TICK
OKAY, OKAY, OKAY!
CLICK CLICK CLICK
YES, MR. PRESIDENT?
JASON, I HAVE A REPORT FROM THE CIA THAT TROUBLES ME!
THIS MULTINATIONAL CORPORATION IS PURSUING THE SPACE SIGNAL ON ITS OWN!! ...I WANT THEM STOPPED!! – SEE TO IT!
YES SIR!
11.

I HAVE AN ACTION MEMO HERE FROM THE PRESIDENT, ... WILL YOU HANDLE IT HOWARD!
DAMN! I GOT THE PRESIDENT'S SOUTHWEST TRIP TO DO BY MONDAY! ... WHICH HAS PRIORITY!? ?
REELECTION, OF COURSE!
MARY-JO PUT GRAYSON ON THIS MEMO FOR ME!
YES, MR. HOWARD
WHY DO I ALWAYS GET THE CRAP JOBS, MARY-JO?
MAYBE IT'S BECAUSE YOU'RE SO SINCERE GRAYSON!
... NEVER MIND HOW THE PRESIDENT FOUND OUT BOSWELL, HE WANTS MULTINATIONAL TO KNOCK OFF THEIR OUTER SPACE DEAL AT ONCE!!
.. C'MON... GRAYSON, I'M JUST THEIR LOBBYIST, I DON'T MAKE POLICY.... I'LL JUST DELIVER THE MESSAGE!
?
RING RING RING RING
RING RING RING RING RING
GRAYSON HERE... OH, HELLO SENATOR! ... SURE, SURE... I'LL BE DOWN IN A MINUTE!!
AH, GRAYSON... NICE OF YOU TO COME OVER SO PROMPTLY. AHHH ABOUT THIS MULTINATIONAL THING THE PRESIDENT WANTS STOPPED ... WE, ER, HAVE A LITTLE PROBLEM HERE... COUGH
12

MULTINATIONAL IS A VERY SUBSTANTIAL CONTRIBUTOR TO OUR PARTY... WE'VE GOT TO LOOK AT THIS IN PERSPECTIVE GRAYSON!
YOU SEE, MY BOY, THIS IS AN ELECTION YEAR... THE PRESIDENT WILL NEED ANY INDUSTRY SUPPORT WE CAN GET HIM!
BESIDES... IN THE LONG VIEW, MULTINATIONAL IS ACTUALLY ACTING IN THE NATIONAL INTEREST!
BUT, SENATOR... IT'S A...AN EXECUTIVE ORDER !!
TUT, TUT, ... GRAYSON, MY BOY, THIS IS NOT A FASCIST STATE! WHY, WE JUST DON'T ORDER A RESPECTED MEMBER OF THE PRIVATE SECTOR AROUND WITHOUT DUE PROCESS!
DO YOU MEAN THAT I'VE GOT TO TELL THE PRESIDENT HE CAN'T STOP THEM?
NOW, GRAYSON... I DIDN'T SAY THAT! SURELY A BRIGHT YOUNG HARVARD MAN LIKE YOU CAN FIGURE A WAY TO COMPLY WITH THE PRESIDENTIAL ORDER AND STILL NOT VIOLATE FREE ENTERPRISE !!
HMMM I SUPPOSE I CAN INSTITUTE A SUIT THROUGH THE ATTORNEY GENERAL'S OFFICE ...HMMM BUT THAT'LL TAKE A YEAR OR TWO TO PROCESS!
GOOD IDEA! ...AND BY THEN A LOT OF THINGS MAY CHANGE! THERE'S AN ELECTION YEAR AHEAD!
... BY THEN YOU MAY BE BACK IN THE PRIVATE SECTOR ...LOOKIN' FOR A JOB - I'M SURE MULTINATIONAL WILL REMEMBER YOUR DEFT HANDLING OF THIS
PERSPECTIVE MY BOY, PERSPECTIVE !!
THANK YOU SENATOR
13.

HELLO HOWARD IT'S GRAYSON HERE .. I AM MOVING ON THAT MULTINATIONAL MATTER THROUGH JUSTICE... OKAY?
RIGHT ...I'LL TELL THE CHIEF IT IS IN MOTION!
CLICK
YAWWN
CLACK
CLACK
CLACK
CLACK
CLACK
CLICK
SCREEE
BAM
?
NADIA ??
LOOKS LIKE I GOT HERE AT THE RIGHT MOMENT ...NO?
14

APPARANTLY YOU'RE IN SOME ONE'S WAY, BLUDD. ...HMMM A HIT MAN -NO IDENTITY CARD! SO, MAYBE THE CIA OR MULTINATIONAL WANT TO SHUT YOU UP... EH?
YOU'RE A SOVIET AGENT NADIA ...WHAT ARE YOU DOING HERE?
I'M NOT HERE AS A SPY... I CAME TO RECRUIT YOU IN BEHALF OF THE UNAFFILIATED PEOPLE... PLEASE, AT LEAST HEAR ME OUT!
THE CONTACT WITH BERNARD MUST BE MADE ON A GLOBAL BASIS... NO ONE NATION HAS THE RIGHT TO DO IT ALONE!
I AGREE!
...THEN, YOU MUST WORK WITH US!
WHO IS US??... SOME CULT GROUP?? ARE YOU ASKING ME TO COMMIT TREASON?
ISN'T THERE A HIGHER RESPONSIBILITY?! SURELY, THERE IS A MORALITY GREATER THAN BLIND LOYALTY TO A POLITICAL ENTITY!!
OH, BLUDD, BLUDD... YOU ARE A SCIENTIST AND A HUMAN BEING THINK, MAN THINK!
...NATIONS ARE CONCERNED WITH NATIONAL WELFARE... WE ARE TALKING ABOUT GLOBAL WELFARE!
15.

WHAT DO YOU SAY... BLUDD ?
WILL YOU? ...WILL YOU?
YES!
I HAVE COME FOR A FAVOR DON CANNELONI...MY SISTER CARLA HAS BEEN MURDERED BY HER HUSBAND VITO LUPO... THE HONOR OF OUR FAMILY HAS BEEN VIOLATED... I CLAIM VENGEANCE!
...YOU GO HOME ROCCO ...WE GONNA TAKE CARE OF IT... DON'T WORRY!
...WE MUST MOVE QUICKLY, NOW, MARCO! AN ADVANCE TEAM MUST GO TO SIDIAMI FOR THE LAUNCHING... WE'LL NEED A VOLUNTEER FOR THE LAB'S FIRST NEW-LIFE-FORM CONVERSION!
HMMMM I'VE GOT A NEW FOLLOWER NAMED VITO LUPO— HE KEEPS ASKING FOR IMMEDIATE TRAVEL INTO SPACE!
STAR PEOPLE
VITO LUPO YOU HAVE BEEN CHOSEN... YOU ARE THE FIRST TO GO ... CONGRATULATIONS !!
STAR PEOPLE
THANK YOU THANK YOU!
16.

BLUDD
LIFE ON ANOTHER PLANET CHAPTER 5
WHAT DO WE KNOW ABOUT JIM BLUDD?
CLASS OF '58 GRAYSON COLLEGE SUMMA CUM LAUDE...TRACK TEAM-VARSITY... DOCTORATE PHYSICS AND THREE BOOKS -14 PAPERS ON CELESTIAL NAVIGATION AND SPACE BIOLOGY
POLITICS ??
NO PARTY AFFILIATION... SIGNED A 'HUMAN RIGHTS' PETITION ONCE...
...RADAR OFFICER IN VIET NAM, THEN TO MT. HOLLY OBSERVATORY UNTIL LAST YEAR WHERE THE CIA RECRUITED HIM FOR THE SPACE SIGNAL THING!
HM-M-M... OKAY CALL NADIA... HAVE HER BRING HIM HERE.
1

RING
RING
RING
RING
RING
RING
NG
RING
HELLO
WE'RE READY ... THE GROUP IS WAITING!
...OKAY, BLUDD! WE'RE ON... THEY'RE WAITING FOR US!
2

DR. PAUL LINUS... DR. ARMAND LIEF... PROFESSOR WILLY GRELL ...RHONDA HENRY— MEET MR. JAMES BLUDD!
HELLO!
PLEASED
SEHER ANGENEMEN
HYA
VE VILL BE BRIEF, MR. BLUDD...VE REPRESENT Z'VORLD ACTION GROUP OF INTERPLANETARY BROTHERHOOD.
WE'RE HORRIFIED AT WHAT'S HAPPENING TODAY!!
PLEASE.... RHONDA...I VILL DO Z'TALK FOR US ALL... NO??
...MR. BLUDD... Z'SPACE SIGNALS HAVE SET IN MOTION A MADNESS THAT VILL END IN GLOBAL DISASTER... AMERICA WALLOWS IN INERTIA... Z'U.S.S.R IS PREPARING ITS OWN LAUNCH...A CRAZY DICTATOR IN SIDIAMI DECLARES HIMSELF A COLONY OF BERNARD... UND MULTINATIONAL IS TAKING MATTERS INTO ITS OWN HANDS... CULTS ABOUND! VE SEE Z'WORLD ORDER...Z'SOCIAL FABRIC COMING APART..
SO, YOU HAVE APPOINTED YOURSELVES GUARDIANS OF GLOBAL SANITY?
IF NOT US... WHO, THEN, WILL REPRESENT THE PEOPLE..?? ...THE MASSES!!
3

WE HAVE PEOPLE IN EVERY COUNTRY ACTIVE IN MOVEMENTS FROM HUMAN RIGHTS TO ENVIRONMENTAL SANITY... THEY'RE SCIENTISTS, THEOLIGANS, ARTISTS DO NOT UNDERESTIMATE OUR INFLUENCE!
WHAT DO YOU WANT OF ME?
VE NEED TIME... VE NEED SOMEONE WHO CAN STOP THAT INSANE LAUNCH!
Y'MEAN THE ONE IN SIDIAMI?
PRE-ZICE-LY!
BY WHATEVER MEANS?
PRE-ZICE-LY!
...YOU ARE HESITANT NEIN?
PRE-ZICE-LY!
DON'T YOU BELIEVE IN THIS CAUSE?
PLEASE... RHONDA! LET HIM THINK!
...OKAY...I DO IT MY WAY... NO INTERFERENCE!
YAH, YAH, YOUR VAY!
4.

ALL PACKED JIM?
JUST ABOUT...
HERE'S YOUR PAPERS... TICKETS... MONEY... YOU'LL BE MET IN SIDIAMI BY HIM... OH, JIM CAN YOU HANDLE THIS...
I'M A SCIENTIST, NADIA, NOT A GUN SLINGER... BUT I THINK I CAN DO WHAT HAS TO BE DONE!
JIM...
...LET'S **HOLD** IT FOR NOW... YOU MAKE EASY ALLIANCES NADIA ... YOU'RE PROBABLY STILL A SOVIET AGENT...
REGARDLESS... I KNOW WHAT I'M GOING TO DO IS RIGHT... MAYBE I'M BEING USED... NO MATTER!
SO, FOR NOW KEEP THE MOTOR RUNNING... WHEN THIS IS OVER WE CAN TAKE UP THE MATTER OF **US**! ... BE SEEING YOU.
5.

THE LAUNCH SITE AND LABS ARE ALL READY IN SIDIAMI... MULTINATIONAL'S BRANCH THERE IS IN CONTROL.
MR. MACREADY, MAY I HAVE A WORD WITH YOU ?!
GET ME GREBE AT STAR PEOPLE
...YES, MR. MACREADY... WE'VE SENT THE FIRST COLONIST FOR CONVERSION... A MR. VITO LUPO - IDEAL SUBJECT! WE'LL NEED MORE FUNDS... OH... AND PLEASE PUSH YOUR ROCHESTER PLANT TO STEP UP THOSE ROCKET COMPONENTS...
STAR
AHEM
GOOD WORK GREBE... I WANT A LAUNCH IN THREE MONTHS!
...NOW WHAT IS IT...??
...WE HAVE A PROBLEM...WASHINGTON REPORTS THE PRESIDENT IS ANGRY AT SIDIAMI'S DECLARATION OF SECESSION FROM EARTH... HE'S NOW DETERMINED TO STOP OUR LAUNCH THERE!
DIDN'T OUR SENATOR BLOCK HIS MOVE ??
IT'S STALLED IN THE JUDICIARY ...BUT THE PRESIDENT IS GOING PUBLIC WITH A DEMAND FOR A U.N. ACTION TO HALT ANY UNILATERAL LAUNCH!
DAMN!
GET DEXTER MILGATE.
HIM?? ... WHY, HE'S A POLITICAL LIABILITY.. AN EX-PRESIDENT A-A LOSER! SHEEE....!!
GET ME DEXTER MILGATE !!!
YES SIR!

STA... PEOPLE
CORA I.. I MUST TALK TO YOU
PLEASE JONESY NOT HERE!
CORA, CORA THIS WHOLE THING IS CRAZY! I NEVER MEANT FOR MY CELL MUTATION THEORY TO BE USED THIS WAY!!
I'M GETTING OUT... I WANT YOU TO COME WITH ME! I LOVE YOU WE CAN...
NO.. NO!
SISTER CORA!
Y-YES MARCO
I'VE GOT GOOD NEWS! WE'RE SENDING 100 FOLLOWERS TO SIDIAMI... AT ONCE
W..WHY?
WE'RE GOING TO BE COLONISTS... AN EARTH COLONY IN SIDIAMI... YOU AND I WILL LEAD THE GROUP!
WELL, HELLO... SIDIAMI... GREAT FLIGHT FOR A HICK AIRLINE EH?
YES
SIDIAIR

7.

I'M HERE LOOKIN' FOR A COUSIN OF MINE, VITO LUPO... WHAT YOU HERE FOR?
BIOLOGICAL RESEARCH
H-E-A-V-Y! I'M IN SPAGHETTI! ...MY COUSIN, LUPO, GOTTA SIGN SOME PAPERS SEE?...YOU GONNA BE HERE LONG?
NO...
AHH MR. BLUDD WELCOME
EXPOR
EVERYTHING'S READY FOR YOU, SIR! MAPS, EQUIPMENT AND A VEHICLE...NADIA GAVE ME THE DETAILS...
THANKS! WHERE'S THE INSTALLATION?
HERE... ABOUT TWO-DAYS DRIVE INTO THE MOUNTAINS
MM
...BEEN LOTS OF ACTIVITY... HEAVY BUILDING CONSTRUCTION STUFF COMING IN ALL THE TIME!
MMM
AND... OH YES... THERE'S AN OVERSEAS CALL FOR YOU...
THANKS
HELLO.. NADIA?? YES, YES NO PROBLEMS.
ER...HYA! SAY, DID YOU SEE THIS GUY COME THROUGH HERE?
HMMM YES... CAME THROUGH HERE LAST WEEK...
WENT UP INTO THE MOUNTAINS --TO THE LAUNCH LAB... WHY?
AH, HE'S MY COUSIN! ...THANKS
8

OKAY, ALL SET... NOW, WHERE'S MY DRIVER ?
...SORRY, MR BLUDD... WE DON'T HAVE ONE... THE ARMY HAS CALLED UP EVERY ABLE BODY... GENERAL AMI'S ORDERS SINCE WE BECAME A COLONY OF...
DAMN
NO SWEAT... I'M A GOOD DRIVER! I'M GOIN' YOUR WAY ANYHOW !!
ROCCO STILLETTO !
... BLUDD! OKAY, YOU'RE ON. WE MOVE OUT TONIGHT!
MILGATE, THIS NATION IS FACING A CRISIS!
CRISIS! I'VE LIVED THROUGH SEVEN OF 'EM BUT, WHEN THE GOING IS TOUGH, THE TOUGH GET GOING!
YES, YES... NOW, WE AT MULTINATIONAL THINK YOU ARE THE MAN TO LEAD US OUT OF IT...
RIGHT! THIS NATION IS FLOUNDERING! THANK HEAVEN I STILL HAVE THE SILENT MAJORITY BEHIND ME...HUMF. I ONLY LOST BECAUSE OF THE PINKO VOTE!
THE COUNTRY IS SOFT SOFT
YES, YES... WE'RE READY TO FINANCE YOUR RETURN... MEDIA EXPOSURE A NATIONAL TOUR ... THE WORKS!
RIGHT I SHALL CAMPAIGN ON THE COMMIE BETRAYAL
WELL... THAT'S NOT QUITE WHAT WE HAVE IN MIND... YOU SEE THE COUNTRY MUST TAKE THE LEAD IN CONTACT WITH THE PLANET BERNARD
HUH ?? BERNARD ?? WHERE'S THAT ?
ITS A DISTANT PLANET THAT SENT OUT A SIGNAL... NOW, UNLESS WE GET THERE FIRST, THIS COUNTRY WILL LOSE WORLD LEADERSHIP!

NOW THE PRESIDENT DOESN'T SEE THE URGENCY... HE'S TALKING ABOUT A JOINT-NATION LONG RANGE PROGRAM!
AAH, IT'S TYPICAL OF THAT FUZZY-BRAINED PINKO LIBERAL CRAP! WE NEED LEADERSHIP
PRECISELY!
I'LL ROUSE THIS NATION
SHOW MR. MILGATE OUT!
MULTINATIONAL IN NEW YORK WANTS A LAUNCH IN 90 DAYS! WHAT'S OUR STATUS?
FUEL JUST CAME IN... THE LAUNCH GANTRY IS UP... IT'LL TAKE A MONTH TO ASSEMBLE THE CAPSULE AND BOOSTER
WHAT ABOUT YOUR 'CARGO', DR. HOAD?
WE'LL BE READY IN 90 DAYS... THE PRIME CULTURE IS INCUBATING - OUR 'FINAL' INGREDIENT IS BEING GROOMED NOW!
WHEN Y'GONNA LAUNCH ME DOCTOR HOAD?
IN NINETY DAYS, MR. LUPO... RELAX, AND BE SURE TO TAKE YOUR PILLS...
MAMA MIA...
DOC... - DOC... I TELL YOU SOMEBODY IS OUT THERE ...HE'S HERE TO KILL ME DO SOMETHIN'
I'LL GET SECURITY
10.

WHAT DID YOU SEE, ROCCO?
I SAW MY COUSIN, VITO... PUFF PUFF PUFF GOTTA SPLIT NOW, BLUDD... GOOD LUCK !!
HEY, WAIT!
ROCCO COME BACK HERE!

WE DON'T KNOW WHO HE IS, GENERAL AMI WE'RE MULTINATIONAL PEOPLE ...OUR JOB IS TO BUILD YOUR LAUNCHING PLATFORM!!... SECURITY IS YOUR AFFAIR!
PROBABLY A U.S. CIA AGENT! PATCH HIM UP OBTUBI AND KEEP HIM HERE UNTIL I DECIDE!! I'VE GOT A DELEGATION TO GREET...
...GET ME DR. HOAD!
OUCH! ...EASY, DOC...
I'M NOT A MEDIC ...I'M A BIOLOGIST! TO THEM ALL 'DOCTORS' ARE THE SAME!
LISTEN...MY NAME IS BLUDD! I'M HERE TO TRY TO STOP THIS MADNESS!
...NAME'S HOAD...SAWYA AT TRIPARTATE!! I'M HERE TO DO A MUTATION EXPERIMENT P-E-R-I-O-D
...AND YOU'RE CRAZY...YOU CAN'T STOP THIS THING NOW!
LISTEN, HOAD... THE CONTACT WITH BERNARD MUST BE A WORLD EFFORT... IT'S THE GREAT CHANCE FOR MANKIND TO UNIFY AT LAST IN AN OUTREACH THAT CAN FINALLY SUBORDINATE NATIONALISM - STAVE OFF WAR AND GIVE CIVILIZED PEOPLE THE FIRST GREAT COMMON GOAL SINCE THE CRUSADES!
LOOK, BLUDD, I AM ONTO THE CREATION OF A NEW LIFE FORM ...MY GOD, WHAT'S YOUR CHEAP POLITICAL CRAP COMPARED TO THAT??
12

HOAD... WAIT... WE'VE GOT TO TALK THIS OUT... LOOK, YOU CAN ALWAYS COMPLETE YOUR EXPERIMENTS ELSEWHERE ...I CAN GET A GRANT
TOO LATE FOR THAT, BLUDD... I'M ALMOST THERE... ALL I NEED IS THE HUMAN FACTOR... WHICH IS ALREADY HERE!
...DR. HOAD... VITO LUPO HAS LEFT HIS ROOM! HE'S RUNNING AROUND...VERY HYSTERICAL!
STRANGE...I'VE HAD HIM ON LIBRIUM FOR A WEEK, HE SHOULDN'T BE HYPER...FIND HIM I'LL TALK TO HIM!
HEY..?! WHAT HAPPENED TO LUPO ?!
I DON'T KNOW DR. HOAD... I... WE JUST FOUND HIM HERE...SHOT BY SOMEONE... MUST HAVE USED A SILENCER... NO ONE HEARD A THING!
OH MAMA MIA
13

IS HE STILL ALIVE ?
BARELY
CAN'T LOSE HIM NOW... WE'VE GOT TO DO THE MUTATION AT ONCE... DAMN IT, HURRY!
OKAY... WE'LL GET HIM READY IN THE LAB...
ZIK ZIK
ZIK ZIK ZIK ZIK
SHEEUT!! THIS IS A GODDAM CHEESE BOX.. C'MON BLUDD, I'M SPRINGIN' YA...
ROCCO!
C'MON, I FOUND AN EMPTY CONSTRUCTION SHACK... WE CAN HOLE UP THERE...
OKAY, ROCCO... WE'RE SAFE FOR A WHILE... NOW, I WANT THE TRUTH! WHAT ARE YOU REALLY DOING HERE ??
AHHH, SHEEUT... I GOT A CONTRACT ON VITO... TRIED TO ICE HIM... BUT I ONLY WINGED HIM, I THINK! GOTTA MAKE SURE... MY PEOPLE DON'T LIKE NO SLOPPY JOBS.... YAWWNN.. I'M SLEEPY...
14.

WELCOME TO SIDIAMI, FIRST EARTH COLONY OF THE PLANET BERNARD...IT IS MY HONOR TO GREET OUR FIRST COLONISTS, THE STAR PEOPLE!
WE ARE HAPPY TO BE HERE, PRESIDENT GENERAL AMI...AND TO ACCEPT YOUR HOSPITALITY UNTIL WE CAN MAKE THE JOURNEY TO BERNARD!
OF COURSE, OF COURSE! MARCO YOU WILL RESIDE IN THE OASIS OF WADIAMI...WHICH YOU CAN SEE FROM HERE...
...AND I SHALL DO YOU THE SPECIAL HONOR OF ALLOWING YOUR YOUNG LADY TO LIVE HERE IN MY PALACE!
OH, NO... THAT'S TOO KIND...!
HEY...NO!... WAIT...SHE STAYS WITH ME!
IT'S NOT WISE TO ARGUE WITH GENERAL AMI, SIR!
HE'S CRAZY, CRAZY!!
PLEASE MARCO... DON'T.. MAKE TROUBLE
AHH...YOUR LEADER, MARCO IS A LITTLE CRAZY NO?...HA, HA, HA,... HA, HA, HA, HA
SEE THAT SHE'S VERY COMFORTABLE, MBESEI!
15

AND...AMURRICA CANNOT AFFORD TO STAND IDLE WHILE THE POTENTIAL TO TAKE WORLD LEADERSHIP SLIPS FROM OUR GRASP... LET THE U.N. TALK!! WE MUST SEND OUR OWN PEOPLE TO BERNARD!
THERE IT IS, MR. PRESIDENT! MILGATE HAS THROWN HIS HAT IN THE RING... WE'RE IN A HELL OF A DOGFIGHT!
WHAT'S YOUR ESTIMATE OF THE SITUATION, JAYSON?
EARLY POLLS SHOW HE'S PICKING UP TREMENDOUS SUPPORT ON THIS ISSUE... THE PARTY 'WHEELS' ARE NOW CONCEDING HIS NOMINATION!
HE'S WRONG!! WRONG... WHAT HE WANTS WILL FORCE RUSSIA'S HAND... IT'LL BRING ON A CONFRONTATION... NO! NO! HE'S A FOOL!
SIR, I MUST ALSO TELL YOU THAT OUR OWN PARTY LEADERS ARE CALLING FOR AN ALTERATION IN YOUR POSITION...
NO!
...WELL, AT LEAST SOFTEN YOUR POSITION UNTIL AFTER ELECTION! ...OUR ANALYSTS FEEL YOU'LL LOSE IF YOU FIGHT MILGATE ON THIS ISSUE!
LOOK, SIR... WHY NOT JUST LET MULTINATIONAL'S LAUNCH IN SIDIAMI GO UP... YOU CAN ALWAYS SAY IT WAS THAT CRAZY AMI - YOU CAN THUS KEEP YOUR OPTIONS OPEN!
WELL... I SUPPOSE I COULD... UNTIL AFTER THE ELECTION
GREAT... CAN WE LEAK IT, SIR.??
OH... GO AHEAD!
...SOURCES CLOSE TO THE WHITE HOUSE REPORTED TODAY THAT THE ADMINISTRATION IS ACCELERATING ITS SPACE PROBE PROGRAM SO THAT THE UNITED STATES CAN RETAIN ITS TRADITIONAL PREEMINENCE IN SPACE EXPLORATION...
16

ABORT
LIFE ON ANOTHER PLANET 6
I CAN'T SIT IN THIS SHACK FOREVER WHILE THEY BUILD-UP THAT LAUNCH!
I'M HERE TO STOP THIS INSANE LAUNCH UNTIL THE NATIONS CAN UNIFY BEHIND THIS PROJECT
AND I NEED YOU TO HELP ME STOP THE LAUNCH IN ANY WAY WE CAN
ANY WAY? ??? SO, BLOW UP THE FRIGGIN' JOINT!
NO, NO, NO! I JUST CAN'T KILL MANY HUNDREDS OF INNOCENT WORKERS!
...THERE MUST BE ANOTHER WAY...
SHEEUT.... I GOT ME A FRIGGIN' NOBLEMAN ...HOO!!
SHEEEEUT!

REJOICE, OH, STAR PEOPLE... IN THE VALLEY BELOW US THEY ARE BUILDING OUR VEHICLE TO ANOTHER PLANET... ONCE THERE WE WILL CREATE A NEW LIFE... FREE OF THE MISERY OF EARTH.
GO... NOW... MY BELOVED PILGRIMS... PREPARE YOURSELVES ... WE DO NOT HAVE LONG TO WAIT!
WILL ALL OF US BE GOING, MARCO?
WHAT ABOUT OUR CHILDREN, MARCO??
TELL US AGAIN, MARCO, WHAT WILL IT BE LIKE?
JONESY... WHAT ARE YOU DOING HERE !?
I'VE COME FOR CORA... I LOVE HER! I MUST SEE HER!
SHE'S A BELIEVER KEEP AWAY FROM HER... SHE'S OURS SHE WILL GO WITH US!
OH, CHRIST, MARCO! Y'R CRAZY! THEY'RE SENDING PLANTS... NOT PEOPLE !!
HA HA HA HA HA HA HA HA HA
WHAT THE HELL IS SO FUNNY?
SHE'S A PRISONER OF GENERAL AMI... DOWN THERE IN HIS PALACE...
OH MY GOD!
I'LL GET HER OUT... HOW DO I GET IN THERE... C'MON... TELL ME...
UP... OVER THAT HILL PAST THE CONSTRUCTION SHACK!
2.

I'VE BEEN THINKING, ROCCO... HOW ABOUT GENERAL AMI ... MAYBE HE CAN BE CONVINCED... TO STALL THE LAUNCH UNTIL...
SHH
POW!!
HOLD IT HOLD !!
WHO ARE YOU ?
JONES... CO-DEVELOPER OF THE CELLULAR MUTATION WITH DR. HOAD!
...I KNOW HOAD!! WHAT ARE YOU DOING HERE, JONES?
I'VE COME TO GET SISTER CORA OUT... SHE'S IN GENERAL AMI'S HANDS!! I'M GOING TO SEE HIM... TALK HIM INTO LETTING HER GO.. I'VE GOT SOMETHING TO TRADE!!
LISTEN JONES, I'M HERE TO DELAY THE LAUNCH... I WANT TO TALK TO AMI TOO!... WHAT DO YOU SAY WE JOIN FORCES?
SURE... WHY NOT?

WELL, WELL SO YOU ARE SISTER CORA!
WHO ARE YOU??
AMI'S No. 1 WIFE...HEARD ABOUT YOU... LOOKIN' YOU OVER!
THE GENERAL HAS PLANS FOR ME...I'LL BE HIS CONSORT IN A NEW...
HA, HA, HA, ...WHAT GAVE YOU THAT IDEA, HA HA!
I KNOW IT... IT'S MY BIG CHANCE... ALL I EVER WAS - A COCKTAIL WAITRESS - NOW I CAN BE SOMEONE WORTHWHILE
YOU WILL BE MY DEAR... YOU WILL BE!
OH, GENERAL, I'LL MAKE YOU A GOOD...
EASY LITTLE ONE!
YOU ARE MY HOSTAGE.... THAT'S ALL! I KEEP MARCO IN LINE WITH YOU...NOW, THAT'S WORTH-WHILE ISN'T IT?
OH, AMI, YOU ARE A CLEVER MAN!
YES, YES YOU'RE RIGHT M'DEAR!
THERE ARE THREE AMERICANS TO SEE YOU GENERAL
4

NOW,... LETME SEE IF I UNDERSTAND YOU.... YOU WANT ME TO DELAY THE LAUNCH... WHY???
AS LONG AS THAT ROCKET IS HERE MULTINATIONAL WILL KEEP FEEDING YOU MONEY... ONCE IT GOES UP, SIDIAMI WILL NO LONGER BE IMPORTANT TO ANYONE!
I AM A COLONY OF THE PLANET BERNARD
AHH BULLSHIT
F'CHRIS SAKE ROCCO!
HA HA HA HA HA HA HA HA
YOU ARE A VERY CLEVER MAN MR. BLUDD... VERY BRAVE! WEREN'T YOU PREVENTED FROM ENTERING OUR LAUNCH SITE?
YES... A BIT ROUGHLY
GET ME DR. HOAD... WE'LL SETTLE THIS NOW!
...THAT'S IMPOSSIBLE GENERAL!! I'VE GOT OUR PASSENGER, VITO READY... WE'RE SET TO GO!
BUT, NOW, HOAD, THESE PEOPLE ARE TRYING TO CONVINCE ME TO DELAY THE LAUNCH
HEY... I WANT VITO...
JONESY!! WHAT THE HELL ARE YOU DOING HERE? I THOUGHT YOU DROPPED OUT!
I WANT CORA
WELL ??
WELL, NOW... PEOPLE WHO WANT SOMETHING ...THAT I UNDERSTAND HAHAHA LET'S DEAL!
5.

CORA, CORA...ITS ME — JONESY... HONEY, THE GENERAL HAS MADE A DEAL!
I'M GOING BACK TO GET MACREADY OF MULTINATIONAL TO GIVE AMI MORE FUNDS... BLUDD IS GOING TO GET THE SOVIETS TO STALL... MARCO IS AMI'S HOSTAGE!
DON'T YOU UNDERSTAND ...WE'RE FREE! YOU CAN COME BACK WITH ME!
PLEASE JONESY... GET UP!
I DON'T WANT TO BE THE WIFE OF A-A-CHEMIST... I WANT SOMETHING MORE... I'VE TASTED A BIT OF IT... NOW I'M GOING TO GET IT!
SOMEHOW
LADIES AND GENTLEMEN THIS HEAH IS A SENATE HEARIN'... NOT A CIRCUS ... PLEASE, COME TO ORDER!
MISTUH MAC READY... CAN YOU EXPLAIN TO US WHY MULTINATIONAL IS UNDERTAKIN' TO FINANCE A LAUNCHING SITE IN SIDIAMI — IN DEFIANCE OF THE STATE DEPARTMENT'S POLICY?
6.

SENATOR,... MULTINATIONAL IS NOT IN DEFIANCE OF ANY FORMAL UNITED STATES POSITION!!... BESIDES, SIDIAMI CLAIMS IT IS A COLONY OF ANOTHER PLANET!
MISTUH MACREADY, THE PRESIDENT IS TRYIN' TO BUILD A INTERNATIONAL COOPERATIVE EFFORT!
IN THAT CASE, DON'T YOU THINK HE SHOULD GET TO THE "PARENT" PLANET FIRST... NOT ITS COLONY?
HA HA HA HA HA HA HA
...YOU SUGGESTIN' SUH, WE MAKE A TREATY WITH A-A PLANET WE HAVEN'T YET VISITED??
...WELL... ISN'T THAT WHAT MY COMPANY IS TRYIN' TO HELP WITH??
SUH, YOUR COMPANY IS TRYIN' TO MAKE FOREIGN POLICY FO' US!
NO SIR... WE ARE ONLY INTERESTED IN SELLING ...KNOW-HOW AND PRODUCT TO SIDIAMI... MAY I MAKE A BRIEF STATEMENT SENATOR?
YES!
HISTORY IS IN THE MAKING HERE! THE FATE OF OUR NATION HANGS ON OUR BEING THE FIRST TO CONTACT THE PLANET BERNARD WHICH HAS SIGNALED US... FOR THIS ADMINISTRATION TO 'DAWDLE' IN A NAIVE EFFORT AT AN INTERNATIONAL EFFORT IS IRRESPONSIBLE!! THIS COUNTRY IS THE MOST ADVANCED ON EARTH... WE RESPECT HUMAN RIGHTS... WE HAVE THE CHRISTIAN ETHIC!! IT IS OUR HISTORIC DUTY TO LEAD - LET THE OTHER NATIONS FOLLOW... THANK YOU!!
BRAVO MR. MAC READY!
NO FURTHER COMMENT, BOYS
IT APPEARS YOU HAVE WON THE DAY, MR. MACREADY
THE PRESIDENT MUST NOW KEEP HANDS OFF SIDI-AMI!
7

WELL, GEORGE... YOU REALLY PULLED IT OFF! THE PAPERS AND TV ARE SINGING YOUR PRAISES THIS MORNING!
MARTHA... DO YOU HAVE TO DRINK SCOTCH AT BREAKFAST?
WELL, WELL I'M FLATTERED! YOU'VE NOTICED. HA... WHAT ELSE CAN A NEGLECTED WIFE DO?
GEORGE, HOWARD CALLED... HE'S GOT A FRIEND, DR. JONES WHO WANTS AN APPOINTMENT... IT'S ABOUT SIDIAMI
JONES ??? DON'T KNOW HIM TELL HIM TO ASK MY OFFICE TO SET ONE UP!
DON'T YOU EVEN WANT TO SEE HOWARD? HE'S YOUR SON!! HE NEEDS A FATHER —TALK TO HIM, GEORGE
THAT 'QUEER' NEEDS A DOCTOR... NOT ME!! GET THE CAR KARL!
..YES SIR
IT'S ALL SET, JONESY ... JUST CALL HIS OFFICE!
THANKS HOWARD!
JONESY... CAN'T YOU STAY A WHILE... WE CAN TALK ABOUT OUR COLLEGE DAYS
SORRY... I..I'M IN A RUSH...'SEE YOU....
DAMN THAT DIRTY DOUBLE CROSSING SONOFOBITCH
...SORRY MR. MILGATE, ...MAC READY WON'T SEE YOU-WE PHONED 5 TIMES!
WELL, WE MIGHT AS WELL FACE UP TO IT, SIR!... YOU'RE OUT OF FUNDS... NOTHING FROM MACREADY NOW THAT HE HAS THE PRESIDENT WHERE HE WANTS HIM!
...BUT WE HAVE ANOTHER POSSIBLE BIG MONEY SOURCE! ... ALBERT APPOLEON - THE PAPER-CLIP MAGNATE! HE'S 'BIG' ON YOU!
OKAY! OKAY! I AM GOING TO GET ELECTED WITHOUT THAT DOUBLECROSSING MACREADY... I DON'T CARE WHO WE GET!
8

MR. MILGATE...IT IS AN HONOR TO BE INVITED TO SUPPORT YOU!
WHY, AS A POOR IMMIGRANT BOY I NEVER DREAMED I WOULD BE UP HERE HELPING TO ELECT A PRESIDENT!
AHEM MR. APPOLEON ... ABOUT FUNDS
SURE, SURE! HOW MUCH Y'GONNA NEED ?
ABOUT 20 MILLION! ...WE'RE BUCKING AN INCUMBENT... WE PLAN TO GO HEAVY ON MEDIA EXPOSURE!
WOW! THAT'S BIG!!.....OKAY, I'LL HAVE TO TAP A FEW FRIENDS...
GREAT! WE'LL WORK OUT HOW YOU FLOW THE FUNDS TO US...LEGALLY
ASK HIM WHAT HE WANTS !!
WANT??...WHY I'M PROUD TO BE ABLE TO BE A PATRIOT...TO REPAY TO THE COUNTRY THAT GAVE A POOR KID HIS CHANCE... I'M A PATRIOT...HELPING HIS COUNTRY IN IT'S HOUR OF NEED...
AHEM
COUGH
SO MAYBE I'LL ASK FOR THE EXECUTIVE WASHROOM KEY IN THE WHITE HOUSE EH? HA HA HA HA!
..THIS WAY SIR
WELL WE'VE GOT IT MR. PRESIDENT !
STOW THAT! WE'RE NOT IN YET!
HMMMM I WONDER ... WHO ARE HIS FRIENDS ??
WHO TH'HELL CARES...DON'T ASK DUMB QUESTIONS HARLICK, JUST GET THINGS MOVING!
9.

OKAY, APPOLEON-YOU GOT FIVE OF THE TOP FAMILIES UP HERE FOR A MEETIN'... IT BETTER BE BIG!
I WON'T WASTE TIME WITH A LOTTA SPEECH... WE GOT A CHANCE TO BUY US A U.S. PRESIDENT!
NOW, THAT IS BIG!
WHO?? ...THAT CREEP-DEXTER MILGATE HAH!!
DON'T KNOCK IT... THIS ISN'T DRUGS OR SKIN! I'M TALKING ABOUT A WHOLE GODDAM ADMINISTRATION IN-OUR-POCKET!
CAAAAMON!
LET HIM TALK
I'M PUTTING UP TWO MILLION-YOU EACH PUT UP 1.8 MILLION... WITH THAT MILGATE'S GOT A SURE SHOT!!
CHEECH
SO, OKAY... HE GETS ELECTED!! WHAT ARE WE BUYIN'??
THE DEAL IS-HE APPOINTS OUR PEOPLE TO KEY POSTS!
AH!
NO ARGUMENT? ... FINE...... OKAY GENTS I WILL DELIVER YOU A PRESIDENT
SO, TONY HOW'S TH' COAST?
SCREWED UP... THIS COUNTRY NEEDS LAW AND ORDER!
HOW'S Y'R KID, JOE?
...GOIN' TO PRINCETON THIS YEAR!
10

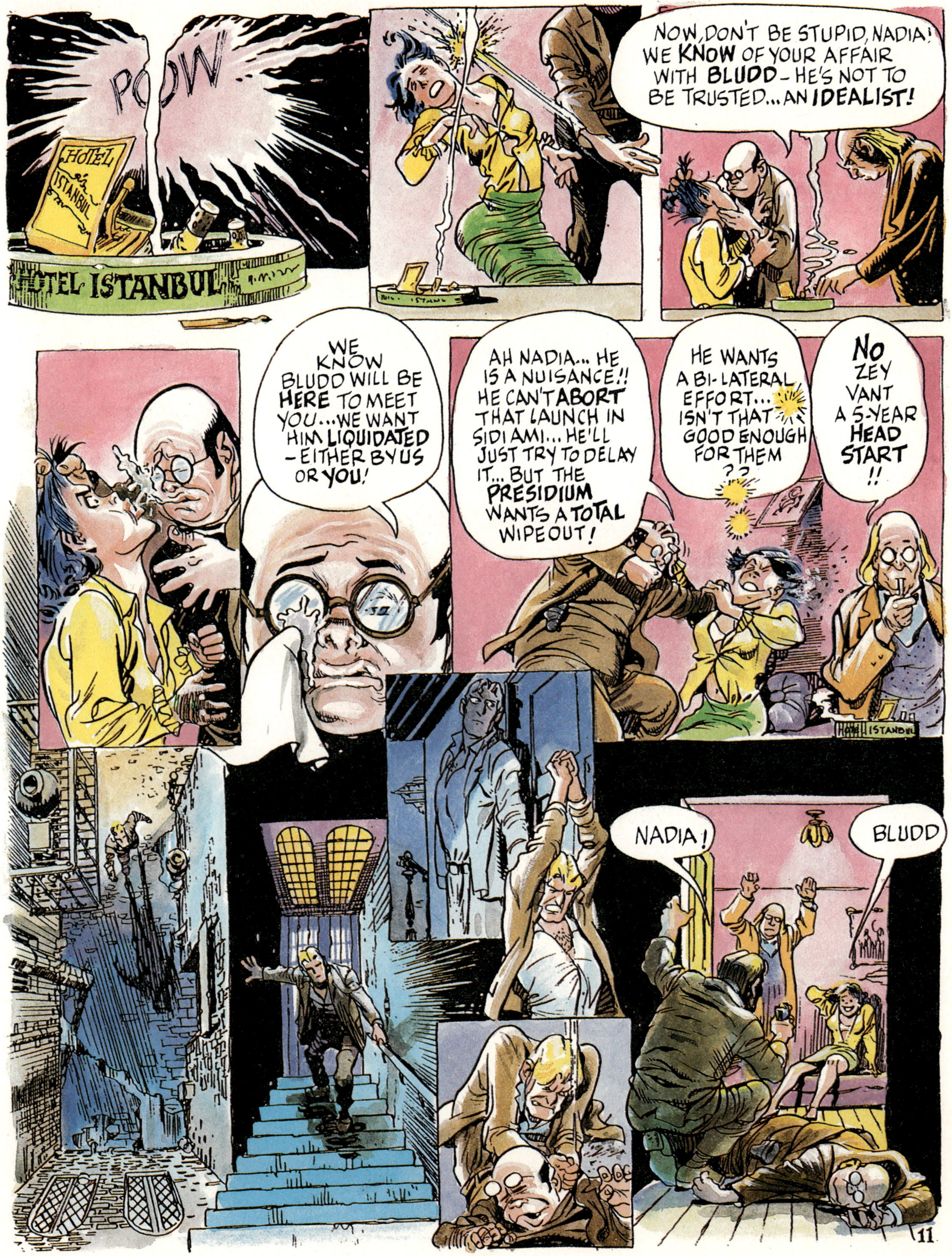
POOW
HOTEL ISTANBUL
NOW, DON'T BE STUPID, NADIA! WE KNOW OF YOUR AFFAIR WITH BLUDD - HE'S NOT TO BE TRUSTED... AN IDEALIST!
WE KNOW BLUDD WILL BE HERE TO MEET YOU... WE WANT HIM LIQUIDATED - EITHER BY US OR YOU!
AH NADIA... HE IS A NUISANCE!! HE CAN'T ABORT THAT LAUNCH IN SIDI AMI... HE'LL JUST TRY TO DELAY IT... BUT THE PRESIDIUM WANTS A TOTAL WIPE OUT!
HE WANTS A BI-LATERAL EFFORT... ISN'T THAT GOOD ENOUGH FOR THEM ??
NO ZEY VANT A 5-YEAR HEAD START !!
NADIA!
BLUDD
11

LISTEN, BLUDD... I AM NOT INVOLVED... IT IS CHUST KGB... HIM... ...YOU ZEE...
SORRY JIM, THEY FOUND OUT ABOUT US!
WHAT DO THEY WANT ...WHY DID THEY HURT YOU?!
THEY WANT YOU OUT! AND THE SIDIAMI LAUNCH ABORTED
MY GOD... GRELL... WHO OWNS YOUR "WORLD CONSCIENCE ORGANIZATION"? IS IT JUST A SOVIET FRONT?... IS IT?
ACH, BLUDD... YOU ARE AN IDEALISTIC FOOL! ... DO YOU THINK ZEY WOULD GO ALONG WITH A JOINT LAUNCH... HOW NAIVE...
GLAK
POM
NO, NADIA
THE ONLY WAY!
POM
OH, JIM... DON'T LOOK SO HURT AND CONFUSED ... IT'S REALLY QUITE SIMPLE! I WILL GO BACK TO RUSSIA AND BLOW THEIR LAUNCH! YOU GO BACK TO SIDIAMI AND STALL THEIR LAUNCH ... MAYBE YOUR DREAM WILL COME TRUE
MEANWHILE WE HAVE 24 HOURS... UNINTERRUPTED!!
12

AT A NEWS CONFERENCE TODAY, THE PRESIDENT ANNOUNCED THAT WHILE THE U.S. CANNOT STOP SIDIAMI FROM MAKING ITS OWN LAUNCH, HE WILL ASK CONGRESS TO PROVIDE FUNDS FOR A U.S. EFFORT AT LAUNCHING A 'VOYAGER' ROCKET TO THE PLANET BERNARD... HOWEVER, HE EMPHASIZED THAT HE STILL HOPES FOR A JOINT-LAUNCH WITH OTHER GREAT NATIONS. SO FAR, HOWEVER, THE USSR VETOED ANY CONSIDERATION OF THE IDEA AT THE U.N! ...THIS IS JAN GATHER IN WASHINGTON...
GOOD EVENING... THIS IS B.B. BLEAKSTONE IN NEW YORK ... WHILE THERE IS GENERAL APATHY IN THE REST OF THE WORLD, THE AMERICAN PUBLIC SEEMS TO REGARD THE PLANET CONTACT AND PROBE AS A KEY POLITICAL ISSUE .. NOW THAT THE PRESIDENT HAS COME OUT WITH AN AFFIRMATIVE POSITION, VOTER SENTIMENT SEEMS TO BE SWINGING HIS WAY... TO DAY, IN KANSAS, DEXTER MILGATE WHO IS LAGGING IN THE POLLS, HAD THIS TO SAY...
...WHEN THE GOING'S TOUGH THE TOUGH GET GOING !!
THIS WAY DR. JONES
THANK YOU, MR. MACREADY
GET MY PLANE READY ...I'M FLYING TO SIDIAMI ...KEEP IT UNDER WRAPS
13

NATIONA AIRPOR
OKAY, BLUDD— HELLO JONES... HOW DID IT GO?
I'VE GOT SOMEONE INSIDE THE USSR TO SABOTAGE THEIR LAUNCH
...AND I GOT TO SEE MACREADY... HE AGREED TO GIVE AMI MONEY WHEN HE FOUND OUT I WAS THE CO-DEVELOPER OF THE 'PLANT PASSENGER'
PLANT PASSENGER ...YES, YES... THAT IS IT... YES THAT'S THE WAY TO STALL IT!
HOW?
SIMPLE, NO PASSENGER NO LAUNCH !!
WOW!...SO WHILE AMI IS BUSY WITH MACREADY WE GO IN AND GET THE 'PLANT'...
SHEEEUT... ALL I WANT IS VITO... LET'S GET MOVING!
YOUR ARRIVAL HERE, MR MACREADY, IS WELL TIMED! ...WE NEED MORE MONEY
GENERAL AMI—OUR STOCKHOLDERS WON'T HOLD STILL FOR BLACKMAIL!
WE HAVE ALREADY GIVEN YOU 50 MILLION IN 'LOAN' TO IMPLEMENT AN EARLY LAUNCH
BUT, MY DEAR MR MACREADY— I'M TORN... THE RUSSIANS, LAST WEEK, OFFERED ME 100 MILLION IF I WOULD DELAY!
SIGH... IT IS NOT EASY TO BE A POOR EMERGING NATION!
14

I DON'T HAVE TIME FOR GAMES GENERAL! TELL ME, FLATOUT.... HOW MUCH WILL IT TAKE TO GET AN EARLY LAUNCH!
WELL... LET ME HAVE 24 HOURS TO DISCUSS IT WITH MY FINANCE MINISTER!
MEANWHILE, WHY DON'T YOU MAKE A LITTLE VACATION OF IT EH?... THE LADIES WILL SEE YOU'RE... HA HA AMUSED... EH?
POM
RESTRICTED AREA
INCUBATOR
YOU COMING WITH US ROCCO?
SHEEEEUT! HE'S OUT COLD! NO...I CAME TO GET VITO, I AINT INNERESTED IN NOTHIN' ELSE... THIS GUY KNOWS WHERE HE IS!
SUIT YOURSELF SO LONG!
CAAAMON WAKE UP!
15.

CUSTOMS
WHEW! WE MADE IT! SAY, I DIDN'T KNOW YOU HAVE A CIA CARD!
USED TO BE... CAME IN HANDY! ... EH??
OPEN UP
OKAY, YOU GUYS! Y'GOT VITO!! HOAD TOLD ME... SO, DON'T GIMME NO CRAP!
YES, ROCCO WE'VE GOT HIM!
VITO WAS... AH... 'MUTATED' SO TO SPEAK... IN A TRUE SENSE THAT PLANT IS VITO..... ...YOU WANT HIM??? OKAY, HE'S YOURS!
SHEEE-UT!
DR. HOAD HAS NEWS, MR MACREADY, IT LOOKS THOUGH THERE WILL BE A DELAY AFTER ALL... THE 'PLANT' 'PASSENGER' HAS BEEN KIDNAPPED BY JONES AND BLUDD!
THE LAUNCH IS POINTLESS SIR, WITHOUT IT!
HOW LONG AND HOW MUCH TO REPLACE IT?
THREE MONTHS!!
AND.. 100 MILLION
16

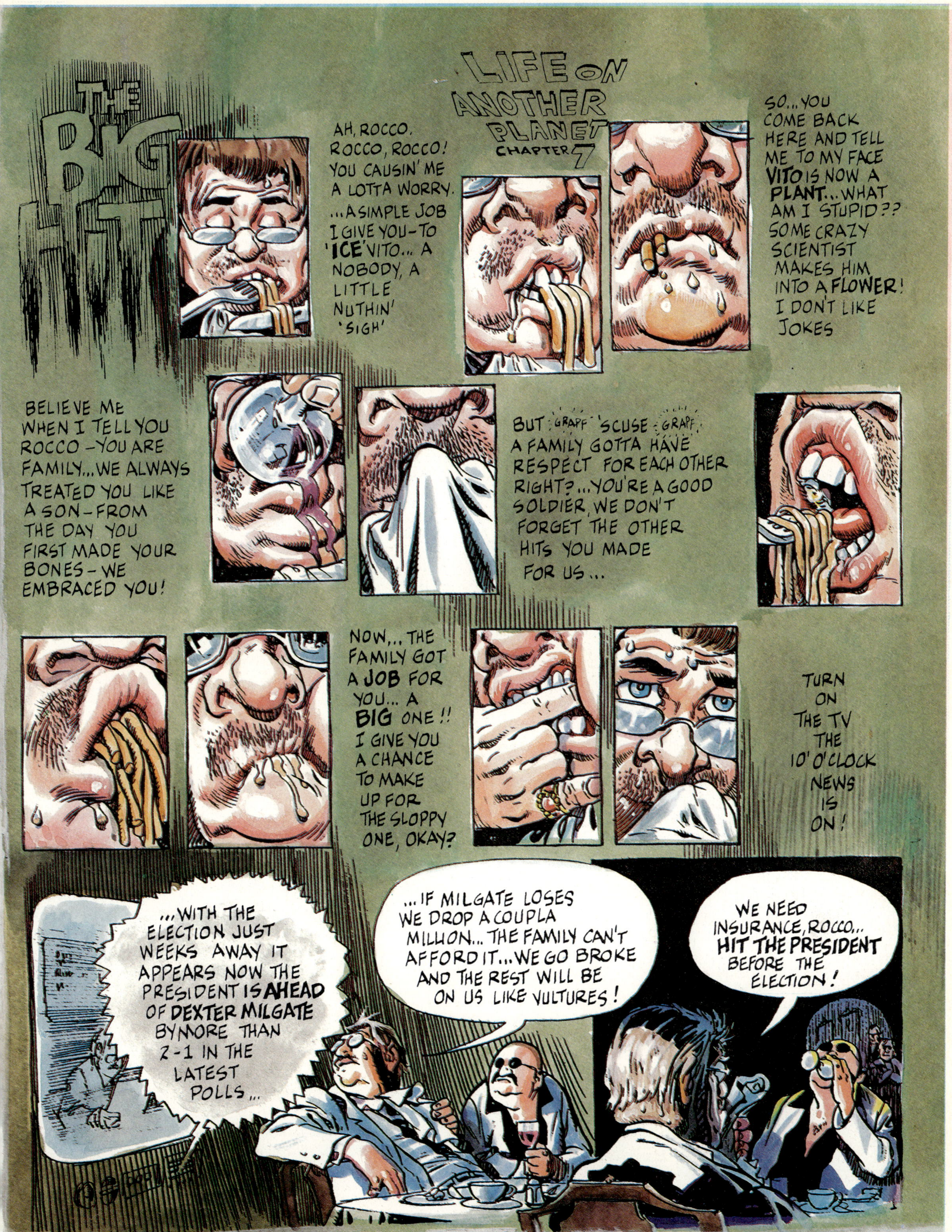
THE BIG HIT
LIFE ON ANOTHER PLANET CHAPTER 7
AH, ROCCO, ROCCO, ROCCO! YOU CAUSIN' ME A LOTTA WORRY. ...A SIMPLE JOB I GIVE YOU—TO 'ICE' VITO... A NOBODY, A LITTLE NUTHIN' 'SIGH'
SO... YOU COME BACK HERE AND TELL ME TO MY FACE VITO IS NOW A PLANT... WHAT AM I STUPID?? SOME CRAZY SCIENTIST MAKES HIM INTO A FLOWER! I DON'T LIKE JOKES
BELIEVE ME WHEN I TELL YOU ROCCO—YOU ARE FAMILY... WE ALWAYS TREATED YOU LIKE A SON—FROM THE DAY YOU FIRST MADE YOUR BONES—WE EMBRACED YOU!
BUT GRAFF 'SCUSE GRAFF A FAMILY GOTTA HAVE RESPECT FOR EACH OTHER RIGHT?... YOU'RE A GOOD SOLDIER, WE DON'T FORGET THE OTHER HITS YOU MADE FOR US...
NOW... THE FAMILY GOT A JOB FOR YOU... A BIG ONE!! I GIVE YOU A CHANCE TO MAKE UP FOR THE SLOPPY ONE, OKAY?
TURN ON THE TV THE 10' O'CLOCK NEWS IS ON!
...WITH THE ELECTION JUST WEEKS AWAY IT APPEARS NOW THE PRESIDENT IS AHEAD OF DEXTER MILGATE BY MORE THAN 2-1 IN THE LATEST POLLS...
...IF MILGATE LOSES WE DROP A COUPLA MILLION... THE FAMILY CAN'T AFFORD IT... WE GO BROKE AND THE REST WILL BE ON US LIKE VULTURES!
WE NEED INSURANCE, ROCCO... HIT THE PRESIDENT BEFORE THE ELECTION!

'HIT' THE PRESIDENT? Y'GOTTA BE CRAZY!
IT'S NOT US, APPOLEON... IT'S THAT CRAZY SOUTH SIDE WOP!
TELL HIM TO LAY OFF!
WON'T DO NO GOOD, HE DON'T LIKE TO LOSE!
MILGATE WON'T AGREE TO THAT!
SO, DON'T TELL HIM?
HELLO HELLO GET MY PILOT HELLO HELLO !!!
SLAM
RATTLE
MULTIAIRS
CLIK
CLIK
DAMMIT GENERAL AMI... WHAT THE HELL'S GOING ON HERE...AM I BEING HELD PRISONER?? WHY IS MY PLANE UNDER GUARD?
HA HA, HA... OH, NO... MR. MACREADY! YOU'RE JUST MY AH GUEST... HA, HA, ... UNTIL YOU'VE MADE UP YOUR MIND ABOUT THE 100 MILLION I NEED... SEE? JUST A GUEST... HA HA HA
2.

... PREPOSTEROUS!! I CAN'T NEGOTIATE A DEAL LIKE THIS SITTING HERE...
NOW, NOW, MACREADY.. YOU'RE ABLE TO DO ANYTHING. .. BESIDES, ONCE Y'R BACK IN THE STATES YOU MIGHT CHANGE YOUR MIND... THEN, THE SOVIETS ARE JUST WAITING TO COME IN ON IT....
AND - IF I HAVE TO DEAL WITH THEM, THEY SEND IN THEIR PEOPLE ... I'LL HAVE A REVOLUTION HERE IN A YEAR AND CHHHK --
WE CAN'T HAVE THAT!
SIR, I HAVE A U.S. WIRE REPORT! .. LOOKS LIKE THE PRESIDENT WILL BE REELECTED! MILGATE IS 'WAY DOWN IN THE POLLS!
DAMN ONCE HE'S IN - HE'LL STOP THE LAUNCH
'SEE... WE HAVEN'T MUCH TIME
AHH, SIR - NEW YORK CALLING ... IT'S YOUR OFFICE
YOUR MOVE MACREADY.
OKAY! YOU GET THE MONEY... I GET THE LAUNCHING... MULTINATIONAL GETS ALL INDUSTRIAL CONCESSIONS!!
YES. YES!
GOOD! .....OH, BY THE WAY YOUR KING IS IN CHECK!
HELLO.. MACREADY HERE! NO, NO, ...FINE...FINE JUST A FEW MORE DAYS...GET OUR FINANCE PEOPLE ON THIS! I WANT A 100 MILLION LOAN TO SIDIAMI, AT ONCE...NO NOT THROUGH THE BANKS..DIRECT!! ALSO, CONTACT OUR SENATOR ... AN' A.I.D.' GRANT AND MILITARY ASSISTANCE! SURE...HE'S ON THE ARMED SERVICES COMMITTEE!! NO, I'LL STAY HERE UNTIL IT'S ALL SET IN MOTION - OF COURSE... THIS IS TOP PRIORITY!!
3

HELLO... ZURICH... MACREADY HERE GET THINGS ROLLING... I WANT YOU TO DIVERT TWO TANKERS WITH PROPELLANT FUEL... NO I DON'T WANT THE STATE DEPARTMENT IN ON THIS... THEY'LL SCREW IT UP... KEEP IT IN COMMERCE... NO, HOLD ALL RELEASES - WE'LL LET IT OUT LATER... AT LAUNCH AS AN AMERICAN EFFORT... SURE, ONCE IT IS UP THEY CAN'T BACK AWAY - SEE IF YOU CAN GET A NAVAL VESSEL AT THE HARBOR IN SIDIAMI
RING
RING
DOCTOR HOAD IS HERE, SIR
BE RIGHT WITH YOU HOAD!
RING
RING
NO, NO... KEEP IT SECRET... WE'RE SCHEDULING A LIFT-OFF IN TWO WEEKS ...YES, BY THEN THE PRESIDENT WILL BE UP TO HIS EYEBALLS IN.... RIGHT, RIGHT.'
...NOW... HOAD... WHAT ABOUT THE 'PASSENGERS'
I TOLD YOU... THREE MONTHS FOR PRIME INCUBATE
NO GOOD! SEND THE CULTURES NOW... THEY CAN INCUBATE DURING THE LONG TRIP!
IT'S TOO RISKY!
DON'T FIGHT ME HOAD... YOU COOPERATE AND I'LL SEE YOU GET A CHAIR ON THE DOME-FETTER FOUNDATION ...JUST MOVE ON IT NOW!
OKAY! OKAY!
WHEW THAT'LL BE ALL FOR TODAY!
MR. MACREADY YOU LOOK TIRED! ...I'VE FIXED A SUITE FOR YOU... IF YOU'LL JUST FOLLOW ME!
?
HAVEN'T WE MET BEFORE ??
YES... YOU'VE GOT A GOOD MEMORY... I'M CORA - YOU SAVED ME ONCE DURING A STREET RIOT - STAR PEOPLE .... REMEMBER ??
4

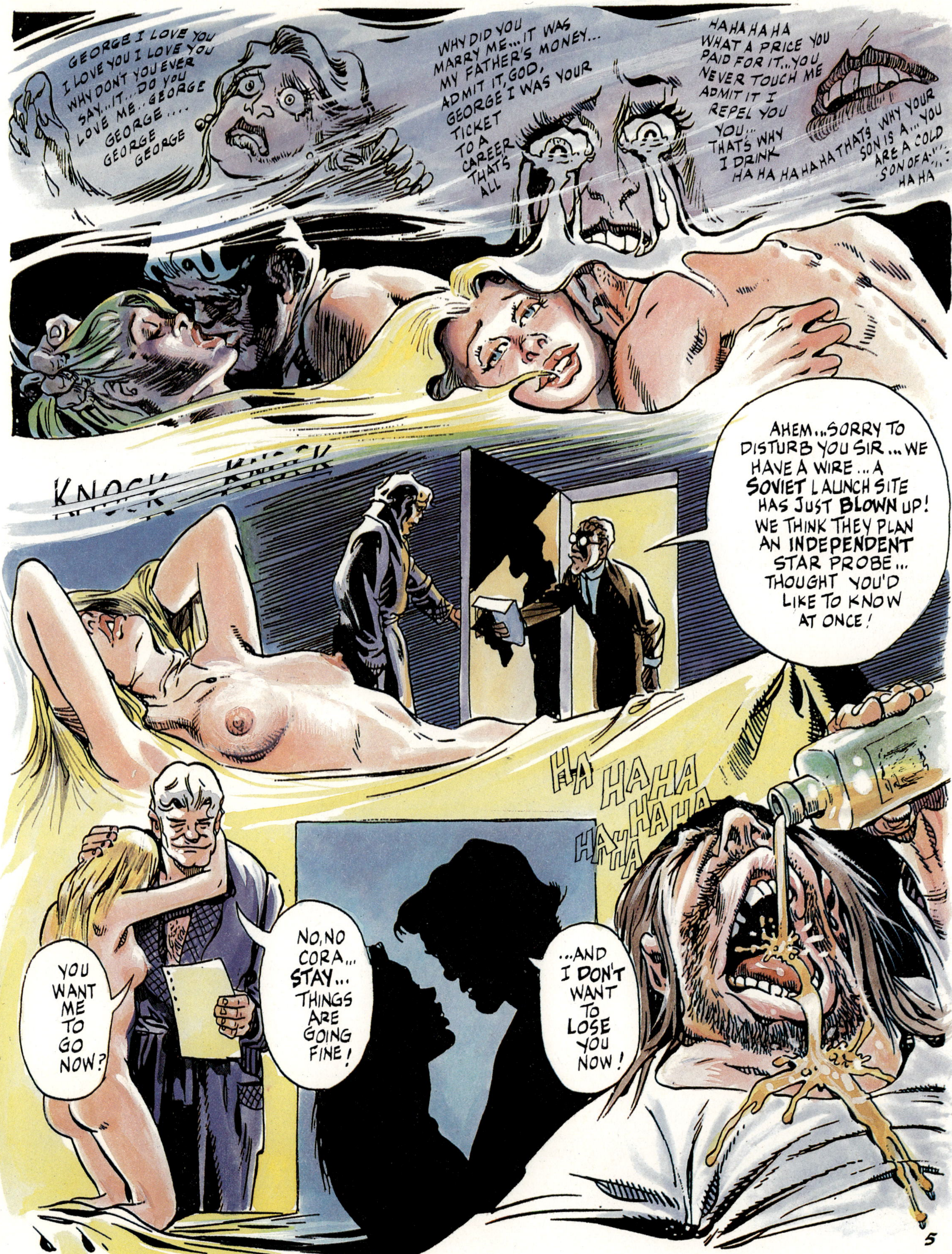
GEORGE I LOVE YOU I LOVE YOU I LOVE YOU WHY DON'T YOU EVER SAY...IT...DO YOU LOVE ME..GEORGE GEORGE.... GEORGE GEORGE
WHY DID YOU MARRY ME...IT WAS MY FATHER'S MONEY... ADMIT IT, GOD, GEORGE I WAS YOUR TICKET TO A CAREER THAT'S ALL
HA HA HA HA WHAT A PRICE YOU PAID FOR IT..YOU NEVER TOUCH ME ADMIT IT I REPEL YOU YOU... THAT'S WHY I DRINK HA HA HA HA HA THAT'S WHY YOUR SON IS A...YOU ARE A COLD SON OF A ·'·'·! HA HA
KNOCK KNOCK
AHEM...SORRY TO DISTURB YOU SIR...WE HAVE A WIRE...A SOVIET LAUNCH SITE HAS JUST BLOWN UP! WE THINK THEY PLAN AN INDEPENDENT STAR PROBE... THOUGHT YOU'D LIKE TO KNOW AT ONCE!
HA HA HA HA HA HA HA HA
YOU WANT ME TO GO NOW?
NO, NO CORA... STAY... THINGS ARE GOING FINE!
...AND I DON'T WANT TO LOSE YOU NOW!
5

LOOK AT THEM THEY'RE WAITING... 100 DREAMS...DREAMS DREAMS, HOAD
HA HA HA HA
MARCO... YOU'RE DRUNK ...LISTEN TO ME... LISTEN!
I ONLY NEED TWO PASSENGERS! ...WE'RE LAUNCHING IN TWO WEEKS... SNAP OUT OF IT!
ONLY TWO? WE GOT A HUNDRED
FOR GOD'S SAKE MARCO Y'LL START A PANIC...LISTEN THIS IS A CELLULAR TRANSPORT GET IT?
WHAT THE HELL DO YOU THINK WE CAME FOR?
WE ALL GO...ALL! I AM THE LEADER OF THE STAR PEOPLE
YOU DON'T UNDERSTAND MARCO! LET ME EXPLAIN...
STAR PEOPLE GREAT NEWS!! WE ARE GOING...NOW, ALL OF US...AS I PROMISED!
NO..NO -DON'T
6.

OH, NO... OH MY GOD MY GOD!
WE HAVE TWO PASSENGERS FOR YOU DR. HOAD!
ER... DR. HOAD IS NOT FEELING VERY WELL... AH... JUST BRING THEM DOWN TO THE LABORATORY! ... HE'LL BE READY FOR 'EM BY... AH, .. MORNING!
CENTRAL INTELLIGENC
THANKS FOR COMING IN BLUDD!
WHAT EVER YOU WANT... THE ANSWER IS NO!
...GUESS YOU KNOW THAT THE SOVIETS HAVE BEEN PLANNING A SECRET LAUNCH TO THE PLANET BERNARD
'FIGURED THEY MIGHT!
...WELL SOMEONE JUST BLEW IT UP.... SABOTAGE!
FROM YOUR SMILE I'D GUESS YOU KNOW WHO DID IT, EH?
WELL... WE KNOW TOO... IT WAS NADIA!
SHE WAS CAUGHT!
...BUT SHE ESCAPED LATER!
SHE'S NOW HOLED-UP IN PARIS.. WE WANT HER! ..SO DO THE SOVIETS... NOW, THE QUESTION IS... WHO GETS TO HER FIRST... I THOUGHT YOU MIGHT.. AH...... BE INTERESTED IN THE JOB, BLUDD!
7

... HOW MUCH TIME DO I HAVE? ... WHAT KIND OF HELP WILL THE CIA GIVE...
I FIGURE YOU HAVE A 48-HOUR HEAD-START ON THEIR KGB... ACTUALLY, SHE CONTACTED US... NADIA WANTS YOU TO COME GET HER... YOU'RE BOOKED ON THE NOON CONCORD!
... HERE'S YOUR RETURN TICKETS... SORRY, THAT'S IT!! WE'RE IN AN ELECTION TIME... NO ONE STICKS HIS NECK OUT... TOO SENSITIVE.. SEE?
YEAH!! THANKS A LOT!
... AND MY FELLOW AMERICANS I CAN ONLY STAND ON MY RECORD
... IN THE FOUR YEARS I HAVE BEEN IN OFFICE WE HAVE LED THIS NATION OUT OF ECONOMIC MESS TO REAL GROWTH IN...
IF I AM RE ELECTED, I PLEDGE TO....
8

ROCCO I DID IT!! LIKE YOU TOL' ME I'M GONNA BE A BIG MAN RIGHT?
9.

PARIS POST
SPECIAL
PRESIDENT SHOT
MILGATE DEPLORES ACT OF MAD ASSASSIN!
PRESIDENT WITHDRAWS FROM ELECTION
ASSASSIN FOUND MURDERED IN ALLEY NEAR SCENE OF CR
FBI STAR INVESTIG
VICE PRES MILGATE CHANCES ELECTION CERTAIN
COMMENT
There is a general consensus among most political observers that the president's withdrawl from the race on the eve of elections will give the election to Dexter Milgate. A late poll of voters shows that the vice president is not so strongly supported and the wave of sentiment favors milgate.
PROMISES A GET TOUGH POLICY
KNOCK KNOCK
..IT'S NADIA ..JIM!
IT'S ALL RIGHT NOW NADIA.. I'LL GET YOU HOME ...IN THE STATES YOU'LL BE SAFE!
DON'T LOOK AT ME JIM..
LET- -- ME... SEE... NADIA!
NO NO PLEAS
OH... GOOD GOD
...WHEN THEY CAUGHT ME, I WAS INTERROGATED FIVE DAYS ... I ESCAPED FROM THE HOSPITAL ... OH, JIM ... DON'T, DON'T LOOK...
10.

CHARLES DE GAULLE AIRPORT
...OUCH!
...YOU OKAY NADIA?
...IT WAS NOTHING! SOMEONE IN THE CROWD JABBED ME WITH HIS UMBRELLA! ...I'M ALRIGHT!
WELL... WE'RE SAFE NOW... ON OUR WAY HOME! ..YOU FEELING OKAY NADIA?
...JUST A BIT SICK... I'LL BE ALRIGHT... PROBABLY JUST A REACTION TO ALL I'VE BEEN THROUGH!
12

WELL... YOU'VE CERTAINLY LIVED UP TO YOUR PROMISE, MR. MACREADY... MY FINANCE MINISTER REPORTS THE MONEY IS ALL CREDITED!
THE LAUNCH PLATFORM IS READY ... LIFT-OFF CAN BE ANY TIME NOW!
HOW'S THE 'PASSENGER' STATUS, DR. HOAD ?
IT'S IN PLACE... THE CULTURE WILL BE ON BOARD... IF ALL MY CALCULATIONS ARE CORRECT, THE PLANTS WILL BE 3-YEARS OLD BY THE TIME THEY GET THERE... THE PLANT INCUBATION IS TIMED TO TAKE 7-YEARS - OR A TOTAL OF 10 YEARS... THE LENGTH OF THE WHOLE JOURNEY TO BERNARD!
WE'VE PROGRAMMED THE ROCKET TO ORBIT THE EARTH TWICE... THEN TO CONTINUE TO DESTINATION BERNARD!
IT WILL ORBIT BERNARD TWICE AND SOFT-LAND THE PASSENGER MODULE... THEN AS THE PLANT'S OWN NERVOUS SYSTEMS REACT TO THE ENVIRONMENT -OR TO ANY OTHER FORM OF LIFE, WE WILL RECEIVE IT HERE! THUS WE WILL BE THE FIRST ON EARTH TO MAKE CONTACT WITH LIFE ON ANOTHER PLANET!
ACTUALLY, THE PLANTS ARE 'WIRED' TO TRANSMITTERS WHICH WILL COMMUNICATE WITH OUR RECEIVERS ON EARTH.
... RIGHT! LOOKING FORWARD TO IT GENERAL... GOOD BY... GOOD LUCK!
SO, MACREADY YOU'LL GET A LIFT-OFF AT YOUR CALL
MULTINATIONAL
13

GOOD BYE! ...DON'T WORRY DR. HOAD! I'LL KEEP MY WORD ON THAT NOBEL CANDIDACY....
TAKE 'ER UP CAPTAIN! I'LL BE GLAD TO LEAVE SIDIAMI ... OH, WHEN WE'RE ALOFT- CALL NEW YORK GIVE THEM OUR E.T.A!
....OH... ER... AH... AND CAPTAIN I DON'T WANT TO BE DISTURBED ...I'M... AH... GOING TO REST DURING THE ENTIRE FLIGHT!
...AND NOW, LITTLE CORA... WHAT ARE WE GOING TO DO ABOUT YOU?.. I'M A MARRIED MAN-YOU KNOW!
I'LL BE YOUR MISTRESS... JUST TO BE WITH YOU... PERHAPS ONE DAY YOU CAN DIVORCE YOUR WIFE... WE CAN MARRY... I SO WANT TO BE SOMEBODY!
YOU WILL DEAR... YOU'LL HAVE IT ALL... I'LL BE GOOD TO YOU!
...AND I'LL BE GOOD FOR YOU!
14

WELL, DOCTOR?
COME HERE BLUDD... WANT TO SHOW YOU SOMETHING!
WE'VE DIAGNOSED THE CAUSE OF NADIA'S ILLNESS! POISON... THE SAME STUFF THAT KILLED DIMYTRI MARJEK IN LONDON TWO YEARS AGO.. KILLS IN 48 HOURS... NO KNOWN ANTI-DOTE!
BUT HOW? SHE WAS WITH ME ALL OF THAT TIME
DON'T KNOW... WE FOUND A TINY PUNCTURE IN HER LEFT BUTTOCK-WE TOOK A SLICE OF TISSUE AND FOUND THIS...
A TINY HOLLOW PELLET WITH A HOLE IN IT... IT CONTAINED THE POISON?
RIGHT! NOW, IN MARJEK'S CASE-IT WAS INJECTED BY SOMEONE WITH AN UMBRELLA ON A LONDON STREET
OH MY GOD!
...WHEN WE WERE BOARDING IN PARIS, SHE SAID SOMEONE JABBED HER WITH AN UMBRELLA
THAT ...WAS IT!
CLICK
CAN I TALK TO HER?
SURE, TRY.... SHE'S SINKING FAST!
NADIA...
15

MILGATE ELECTED LANDSLIDE
MILGATE WINS ELECTION BIG
WASHINGTON DAILY
MILGATE VOWS U.S. WILL LEAD IN SPACE PROBE
PRESIDENT -ELECT WILL STRONG
BELIEVES THERE IS LIFE ON ANOTHER PLANET U.S. WILL BE FIRST TO MAKE CONTACT. WARNS OTHER NATIONS WILL NOT TOLER
MILGATE RECOGNIZ
SOCIETY
MACREADYS SEPARATE
WEALTHY TYCOON IS
FILING FOR DIVORCE
SOCIAL SET STUNNED
SEPARATED
16

THE LAST CHAPTER
NOTES FOR A MONOGRAPH ENTITLED INTELLIGENT LIFE ON ANOTHER PLANET
By James Bludd
It is Now six Months since the election of Dexter Milgate
ITEM: George Macready, Chairman of Multinational has been appointed 'special advisor to President Milgate. Appoleon is chief of the White House staff.
ITEM: The launch in Sidiami has been delayed to give the U.S. a chance to Consolidate its own Control and identification with this probe. It is now viewed as a U.S. Undertaking!
ITEM: The Soviet protests in the U.N. go unheeded.
...WELL, HYA, CORA! ...WANGLED AN INVITATION SO I CAN SEE HOW THE NEW MRS. MACREADY MOVES IN BIG-TIME SOCIETY!
PLEASE JONESY... GO AWAY!! I'M HAPPY ... ALL MY DREAMS HAVE COME TRUE. GEORGE IS GOOD TO ME... LET ME ALONE
... CAN WE TALK OUT HERE APPOLEON
...ER, SURE, MARIO
DAHLING,.. YOU LOOK LOVELY
AH, THANK YOU DEAR...OH, YOU KNOW, WASHINGTON COCKTAIL PARTIES... HARDLY A CASUAL AFFAIR.
I KNOW, HASTINGS, I KNOW... OH, 'SPLINT' - CAN I TALK WITH YOU?
MACREADY... THE SITUATION IS HEATING UP... SOVIETS HAVE CUBAN TROOPS ON THE SIDIAMI BORDER ... CIA PREDICTS A COUP IMMINENT... MILGATE CAN'T STALL MUCH LONGER... YOU'RE HIS CLOSEST ADVISOR... TALK TO HIM!! HE MUST MOVE NOW... I CAN'T RUN THE STATE DEPARTMENT BY EAR, Y'KNOW!!
SURE MAC'... EXCUSE ME, SAM!

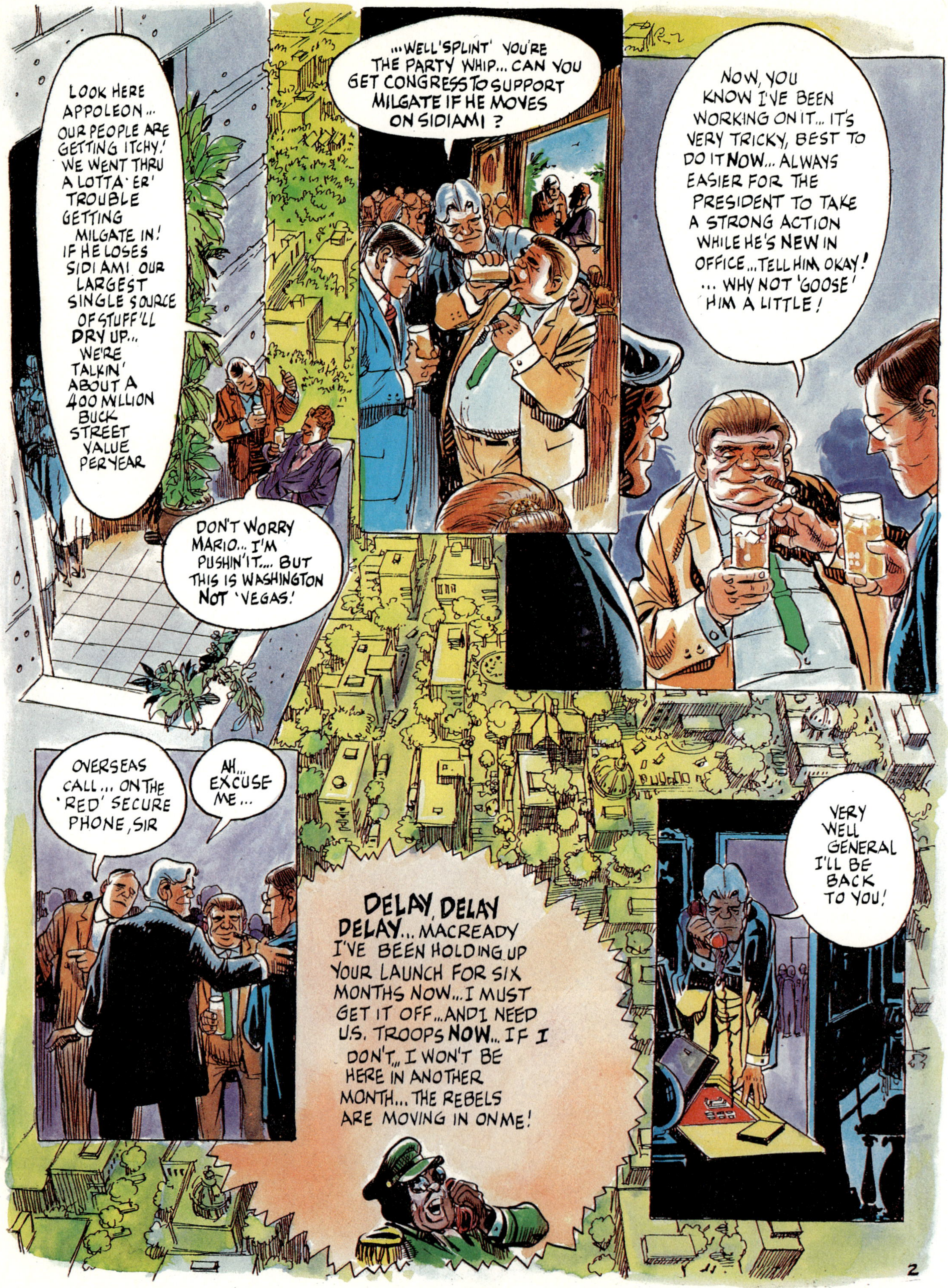
LOOK HERE APPOLEON... OUR PEOPLE ARE GETTING ITCHY! WE WENT THRU A LOTTA 'ER' TROUBLE GETTING MILGATE IN! IF HE LOSES SIDIAMI OUR LARGEST SINGLE SOURCE OF STUFF'LL DRY UP... WE'RE TALKIN' ABOUT A 400 MILLION BUCK STREET VALUE PER YEAR
DON'T WORRY MARIO... I'M PUSHIN' IT.... BUT THIS IS WASHINGTON NOT 'VEGAS.'
...WELL 'SPLINT' YOU'RE THE PARTY WHIP... CAN YOU GET CONGRESS TO SUPPORT MILGATE IF HE MOVES ON SIDIAMI ?
NOW, YOU KNOW I'VE BEEN WORKING ON IT... IT'S VERY TRICKY, BEST TO DO IT NOW... ALWAYS EASIER FOR THE PRESIDENT TO TAKE A STRONG ACTION WHILE HE'S NEW IN OFFICE... TELL HIM OKAY! ... WHY NOT 'GOOSE' HIM A LITTLE!
OVERSEAS CALL... ON THE 'RED' SECURE PHONE, SIR
AH... EXCUSE ME...
DELAY, DELAY DELAY... MACREADY I'VE BEEN HOLDING UP YOUR LAUNCH FOR SIX MONTHS NOW... I MUST GET IT OFF... AND I NEED U.S. TROOPS NOW... IF I DON'T... I WON'T BE HERE IN ANOTHER MONTH... THE REBELS ARE MOVING IN ON ME!
VERY WELL GENERAL I'LL BE BACK TO YOU!
2

KARL, GET ME THE PRESIDENT
YES SIR
MR. PRESIDENT YOU MUST MOVE NOW! ...YES, THEY WILL SUPPORT YOU.... YES,....YES....
THANK YOU MR. PRESIDENT
HELLO... GENERAL AMI...... MACREADY HERE
LAUNCH NOW !!
WELL, GENTLEMEN IT'S DONE... THE BALOON IS UP!
I NEED ANOTHER DRINK
3

THIS IS SAM GATHER IN SIDI, THE CAPITAL OF SIDIAMI WHERE THE EARTH'S FIRST SPACE LAUNCH TO CONTACT AN INTELLIGENT LIFE ON ANOTHER PLANET HAS JUST TAKEN OFF! THERE IS AN EERIE CALM HERE... EXCEPT FOR THE PRESENCE OF U.S. NAVAL VESSELS, THERE IS NOTHING TO INDICATE THAT THIS WAS THE 'EVENT' OF OUR CENTURY! ... BACK TO BGS IN NEW YORK.... THE REACTION TO THE 'EVENT' HERE IS INCREDIBLE IN THAT THERE IS LITTLE OR NO REACTION... THE PUBLIC SEEMS UNCONCERNED!! OH, THERE WAS A MILD FLURRY OF SPACE TECHNOLOGY STOCKS ON WALL STREET TODAY...
COMRADE TOLKININ... EXCUSE ME I HAVE A CODE DISPATCH FROM MOSCOW...
GET ME AN APPOINTMENT WITH THE PRESIDENT I MUST TALK TO NO ONE ELSE!
CLICK
MR. PRESIDENT... THE SOVIET AMBASSADOR IS ASKING FOR AN IMMEDIATE AUDIENCE... HE SAYS MOSCOW WANTS 'CLARIFICATION' OF YOUR POSITION....
SCREW HIM!! ...OUR POSITION IS ...WE DON'T HAVE TO DISCUSS ANYTHING WITH THEM! THE LAUNCH IS UP... P-E-R-I-O-D... TELL THEM TO GET THEIR GODDAM CUBANS OUT OF SIDIAMI FIRST... THIS IS A U.S. SHOW... IF THEY WANT TO GO TO WAR OVER THIS.. FINE..I'M READY... WHEN THE GOING GETS TOUGH THE TOUGH GET GOING!!
4

HOTEL
HE'S IN ROOM 48 -FOURTH- FLOOR SIR
GOOD WORK... THANK YOU
MACREADY!? WHAT ON EARTH DO YOU WANT WITH ME??
WELL, BLUDD... WE HAD QUITE A TIME FINDING YOU! ... UNDERSTAND YOU'VE BEEN GRIEVING OVER A LOSS ... MISS NADIA, EH?... ... SORRY...
5

FIRST, LET ME UPDATE YOU! OUR SIDI SPACE LAUNCH WAS INTERCEPTED AND DESTROYED BY THE SOVIETS ... PROBABLY FROM THEIR SOYEZ SPACE PLATFORM...
HA!! GREAT NEWS!
8000 CUBAN TROOPS HAVE JOINED A REBEL TRIBE TO OUST GENERAL AMI! ... THEY'RE MARCHING ON OUR LAUNCH SITE IN SIDI...
REALLY?
ER... WHAT TIME IS IT... GOT A RADIO?
HERE
... TODAY, AT 5AM NEW YORK TIME CUBAN TROOPS, SUPPORTED BY ELEMENTS FROM THE SOVIET UNION HAVE OVERRUN OUR LAUNCH SITE IN SIDI... IN THIS ACTION 500 OF OUR MARINE FORCE THERE TO PROTECT AMERICAN INTERESTS WERE KILLED OR WOUNDED...
.... THE PRESIDENT HAS ASKED FOR AN EMERGENCY SESSION OF THE U.N. TO CONSIDER SUITABLE ACTION... IN LATER DEVELOPMENTS PRESIDENT MILGATE CALLING THIS A GRAVE NATIONAL CRISIS HAS ASKED FOR A JOINT SESSION OF CONGRESS TO SEEK WAR-EMERGENCY POWERS....
DAMN! MACREADY... YOU FINALLY DID IT! ... EXACTLY AS I FEARED!! WE'RE BEING LED INTO A WORLD WAR!
NOT YET, NOT YET
NOW WHAT?
MILGATE WILL, OF COURSE, DESTROY THE SIDI LAUNCH SITE - TAKE IT OUT OF CONTENTION! ... IT'LL NEUTRALIZE THE SOVIET CLAIMS!
CLIC
6

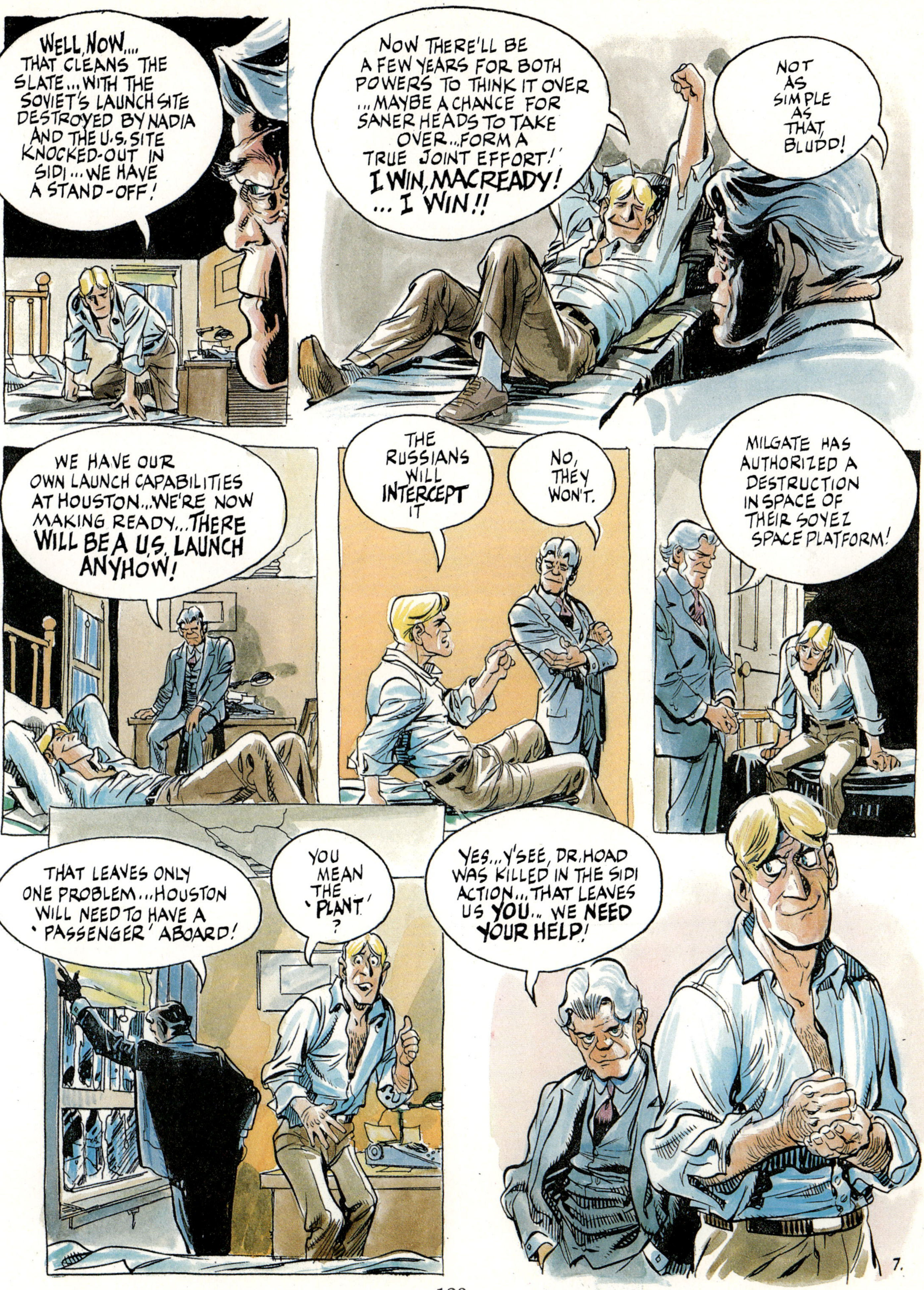

WELL, NOW.... THAT CLEANS THE SLATE... WITH THE SOVIET'S LAUNCH SITE DESTROYED BY NADIA AND THE U.S. SITE KNOCKED-OUT IN SIDI... WE HAVE A STAND-OFF!
NOW THERE'LL BE A FEW YEARS FOR BOTH POWERS TO THINK IT OVER ... MAYBE A CHANCE FOR SANER HEADS TO TAKE OVER... FORM A TRUE JOINT EFFORT!' I WIN, MACREADY! ... I WIN!!
NOT AS SIMPLE AS THAT, BLUDD!
WE HAVE OUR OWN LAUNCH CAPABILITIES AT HOUSTON... WE'RE NOW MAKING READY... THERE WILL BE A U.S. LAUNCH ANYHOW!
THE RUSSIANS WILL INTERCEPT IT
NO, THEY WON'T.
MILGATE HAS AUTHORIZED A DESTRUCTION IN SPACE OF THEIR SOYEZ SPACE PLATFORM!
THAT LEAVES ONLY ONE PROBLEM... HOUSTON WILL NEED TO HAVE A 'PASSENGER' ABOARD!
YOU MEAN THE 'PLANT'?
YES... Y'SEE, DR. HOAD WAS KILLED IN THE SIDI ACTION... THAT LEAVES US YOU... WE NEED YOUR HELP!
7.

I DON'T HAVE THE 'PLANT'... I GAVE IT AWAY!
THEN YOU MUST GET IT BACK!
YOU TOOK IT FROM SIDI...YOU KNOW WHO HAS IT... I MUST HAVE IT ON BOARD!
WELL, WELL, SO THIS HAS TAKEN ON SOME KIND OF MEANING TO YOU... EH, MACREADY?!!
I DON'T NEED PSYCHOLOGY BLUDD... I NEED YOUR HELP!
HOW MUCH DO YOU WANT BLUDD? MONEY?? NAME IT!
I WANT A PERMANENT POST... SOMEWHERE... WHERE I'LL BE OUT OF THIS ROTTEN WORLD... GET ME A POSITION IN THE MT. HEAD SPACE OBSERVATORY... AS A MONITOR IN THE SPACE RADIO EMISSION LAB... I WANT TO BE UP THERE WHEN THIS WHOLE MESS YOU STARTED ERUPTS INTO AN ARMAGEDDON!
... DONE!!! YOU'VE GOT IT JUST GET ME THAT 'PLANT', BLUDD!
DAILY GLOBE
U.S. LAUNCHING OWN STAR PROBE
DESTRUCTION OF SIDI
WILLNOT STOP U.S.
SAYS MILGATE
YANKS WIN

BELCH
AH VITO VITO YOU LOOK SO NICE TODAY
OH ROCCO ROCCO... HE ANSWERED ME ...SEE HOW THE LEAF TWITCHES AHHH
SHEEUT!! YER NUTS, ROSA!
YOU GAVE HIM TO ME... JUST THINK, MY POOR DEAD SISTER, CARLA'S HUSBAND...A..A PLANT!? YOU KNEW THAT HIM-AND-ME:..I LOVED HIM! SO YOU BROUGHT HIM TO ME...TO ME, ROSA THE OLD MAID...
SO, I LET YOU STAY AND LIVE WITH ME... NOW I GOT EVERYTHING ... I GOT VITO AND I GOT YOU!
BELCH
ROCCO... ROCCO, ROCCO... WE ARE LIVING TOGETHER -- LIKE MAN AND WIFE .. SO, WHY WE DON'T ENJOY IT ..EH, EH, EH...?!
CAAAMON ROCCO
LEGGO... GODAMMIT!
GEDAAADA HERE Y'FATFUKKIN COW... I DON'T WANNA SKREW!!

I JUST AINT INNERESTED! YOU.. YOU UNDERSTAND?? I CAN'T... I DON'T LIKE IT WITH WIMMEN... SEE?!
I'M GONNA TELL MY UNCLE, DON CANNELONI ON YOU ...HE'LL FIX YOU... YOU GODDAM FAIRY...
HAW!! WHY THE HELL DO YOU THINK I'M LIVIN' HERE?! Y'R UNCLE OWES ME... HE SENT ME HERE 'TILL THE HEAT'S OFF! SO, HE AINT GONNA DO NOTHIN' ... BESIDES, I KNOW TOO MUCH! NOW, SHADDAP!
BELCH
OH VITO, VITO, VITO ...YOU ALL I GOT NOW BABY...
AHH JONESY... YOU DEVIL.. WHO IS THIS... HMM HANDSOME ♫
IT'S NOT LIKE THAT HOWARD... LISTEN, HE'S JIM BLUDD... A-A SCIENTIST.... WE NEED YOUR HELP!
WE'RE LOOKING FOR A 'HIT' MAN... ALL WE HAVE IS: ...HE LOOKS LIKE THIS... AND HIS NAME'S ROCCO.
REALLY... THE ONLY TIME I EVER SEE YOU IS WHEN YOU WANT A FAVOR ... THIS ISN'T CONNECTED WITH MY FATHER...?' WHY ME?
ALL THE C.I.A HAS ON HIM IS - HE FREQUENTS GAY BARS... YOU'RE INTO THAT SCENE ...HELP US FIND HIM... PLEASE... HOWARD...
ROCCO, PLEASE
COPS ?
NO, NO, I'M JAMES BLUDD
LET HIM IN...
10.

I CAME FOR THAT PLANT ROCCO!
HA, HA, HA SHEEUT... SURE, TAKE IT... TAKE THE FRIGGIN' THING!
No, No, No! MAMMAMIA ROCCO... IT'S MINE... IT'S MY VITO!
LEGGO CARLA... I'M TIRED WATCHIN' YOU SLOBBER OVER IT... L-E-G-G-O
ROCCO, NO DON'T TAKE MY VITO!
OH, JONESY, ... MR BLUDD I CAN'T STOP THE BLEEDING ... WE MUST GET HIM TO A HOSPITAL
YES.. IN A LITTLE WHILE HOWARD... NOW JONESY, CAN YOU HOLD ON LONG ENOUGH TO HELP ME...
I'LL TRY ... WHA ??
I'M GOING TO IMPLANT AN EXPLOSIVE ...ONE THAT CAN BE SET OFF BY A RADIO SIGNAL FROM EARTH
GOOD GOD BLUDD Y'CAN'T
IT'S MY LAST CARD JONESY !!
11

I CAN GIVE THE WORLD ONE LAST GIFT OF TIME...
IF THE PLANT LANDS ON BERNARD... AS A U.S. PROPERTY... WE'LL HAVE ANOTHER WAR FOR CONTROL...
IF, HOWEVER THE WORLD HAS MORE TIME TO DISCUSS IT
A JOINT-EFFORT, --A COMPROMISE WILL EMERGE... TIME, JONESY, TIME....
YOU'RE A FANATIC BUDD
MAYBE... IT'S WHAT OFTEN BECOMES OF A FRUSTRATED IDEALIST... THERE, NOW...
I... I... CAN'T
HOWARD... GET HIM TO A HOSPITAL ...I'VE GOT TO DELIVER THIS PLANT...
EMERGENCY
HOWARD... HOW IS JONESY?
HE... DIED AN HOUR AGO...
12

WHATS WITH THAT GUY, BLUDD..? BEEN HERE SOME TIME NOW... A LONER! ... JUST SITS AT THE MONITOR... WRITES IN HIS DIARY AND HANGS ONTO THAT BLACK BOX... STRANGE...
YEAH... I PEEKED INTO THAT BOX... NOTHING IN IT BUT A RADIO SENDING DEVICE... IT COULD SEND A LONG-DISTANCE SIGNAL ... MAYBE A BEEP
WHAT EVER FOR ?
WHO KNOWS ?? MAYBE HE'S A 'NUT!'
WELL, HE CAME WITH 'TOP' WASHINGTON PROTECTION..... ...NONE OF OUR BUSINESS..
THIS IS THE NINE O'CLOCK ♪ NEWS ♪
... AND THE SPACE CENTER AT HOUSTON HAS ANNOUNCED THAT A SPACE PROBE CONTAINING A 'MONITOR' PLANT HAS BEEN LAUNCHED... IT WILL REACH THAT DISTANT PLANET IN 10 YEARS WHERE HOPEFULLY WE CAN ESTABLISH CONTACT WITH AN INTELLIGENT LIFE FORM THAT HAS BEAMED A SIGNAL TO EARTH...
13

AND NOW THE LATE BALL SCORES
CLICK
14

WE INTERRUPT THIS PROGRAM FOR A SPECIAL BULLETIN
NASA REPORTS IT HAS LOST ALL CONTACT WITH THE BERNARD SPACE PROBE...IT IS BELIEVED THAT A MALFUNCTION MAY HAVE CAUSED AN EXPLOSION IN THE CARGO AREA AND
CLICK
15

MR. BLUDD... WE'VE JUST TAPED A RADIO SIGNAL FROM 'WAY OUT IN SPACE
STRANGE... IT IS DIFFERENT FROM ANY WE'VE EVER HAD... SEEMS TO BE ORGANIZED INTO INTELLIGENT GROUPINGS.. SEE WHAT Y'CAN MAKE OF IT!
HMM... REALLY?
THE LAST CHAPTER